C.R. Bridges · A. Curtis · D. Sanders (Eds.)

Experimental Biology Online · Annual 1998

Springer

Berlin
Heidelberg
New York
Barcelona
Hong Kong
London
Milan
Paris
Singapore
Tokyo

EBO

Experimental Biology Online Annual 1998

C.R. Bridges · A. Curtis · D. Sanders

CD-ROM and print archive edition of the journal „Experimental Biology Online“ 1998

Springer

Professor Christopher R. Bridges
Institut für Zoophysiologie
Lehrstuhl für
Stoffwechselphysiologie
Heinrich-Heine-Universität
D-40225 Düsseldorf
Germany

Professor Dale Sanders
Department of Biology
University of York
Heslington
York YO1 5DD
United Kingdom

Professor Adam Curtis
Division of Infection and
Immunity
Joseph Black Building
University of Glasgow
Glasgow G12 8QQ
United Kingdom

Additional material to this book can be downloaded from http://extras.springer.com.

ISSN 1435-3237
ISBN-13: 978-3-642-64248-7 Springer-Verlag Berlin Heidelberg New York

Cataloging-in-Publication Data applied for

Die Deutsche Bibliothek - CIP-Einheitsaufnahme
[Experimental biology online / Elektronische Ausgabe]
Experimental biology online ; EBO ; annual ... [Elektronische Ausgabe] [Computerdatei]. - Berlin ; Heidelberg ; New York ; Barcelona ; Budapest ; Hong Kong ; London ; Mailand ; Paris ; Singapur ; Tokio : Springer
Titel auf dem Label
Gedruckte Ausg. u.d.T.: Experimental biology online
1998 (1999)
ISBN-13: 978-3-642-64248-7 e-ISBN-13: 978-3-642-60083-8
DOI: 10.1007/978-3-642-60083-8

Softcover reprint of the hardcover 1st edition 1999

Cover design: design & production GmbH, Heidelberg
Production: Springer-Verlag, Thomas Bastuck, Journal Production Department II, Heidelberg
SPIN 10708618 31/3190 . 5 4 3 2 1 0 . Printed on acid-free paper

Aims and Scope - Call for papers

Experimental Biology Online - EBO is a new, multimedia electronic journal sponsored by three international biological societies with a membership of over 6,000 worldwide.

Its aim is to provide a medium for rapid, readable, refereed and rigorous short, topical research papers in the general field of experimental biology in its broadest sense, ranging from the molecular and cellular boundaries in cell and plant biology to the integrative study of whole organism including the experimental analysis of behaviour. It is organized in three sections – Animal Biology, Cell Biology and Plant Biology.

All papers are submitted and refereed by peer-review electronically either via e-mail or anonymous FTP server. This usage of the Internet at every step of the reviewing and publication process leads to exceedingly short publication times of a few weeks only from submission to appearance. The journal is accessible on the World Wide Web, with long-term access guaranteed.

How to submit a manuscript?

Papers may be submitted directly via anonymous ftp to science.springer.de/ebo/ and to the appropriate directory:

animal/
cell/
plant/

Please choose unambiguous file names (e.g. starting with your initials and the date *DDMM*). For those who are unfamiliar with ftp clients we have provided detailed descriptions of this procedure for PCs running MS Windows and for Macintosh computers. Alternatively, files may be sent on diskette(s) including a similar description to:

Thomas Bastuck
Journal Production Department II
Springer-Verlag
Postfach 10 52 80
D-69042 Heidelberg
Germany

Tell us you have sent a manuscript!

Finally, fill in the checklist and send it to Springer-Verlag by clicking on the submit button.

You may also pre-submit your manuscript title and a short description of the paper before you send the final version. This allows the editors to contact prospective referees before submission and speed up the refereeing process.

Receipt of your manuscript will be acknowledged by the editor, marked up as HTML by Springer, and placed in a password-protected area of the Springer WWW site. You will be sent a copy of the marked up HTML format via e-mail to check for errors in mounting. Referees are provided by e-mail with the password to read the paper online with their Web browser. They communicate by e-mail with the editor, who in turn communicates by e-mail with you.

If the manuscript is accepted with modifications you simply resubmit the revised version, taking into account the referees comments, the files are marked up in HTML, copy edited and then transferred to the public portion of the archive and added to the contents list for that volume. Thus, this could be the quickest paper you have ever published!

For **further technical details** pertaining to file formats please visit EBO's homepage at:

http://link.springer.de/journals/ebo/

Subscriptions can either be placed via your bookseller agency, or sent directly to:

Springer-Verlag
Karin Tiks
Heidelberger Platz 3
D-14197 Berlin, Germany
Tel. ++49-30-82787-358
FAX ++49-30-82787-448

e-mail: *subscriptions@springer.de*

1999 EBO subscription rates

Institutional subscription: DM 238.- (US $ 143)
Personal subscription: DM 98.-
subscription for members of affiliated societies (SEB, SICB, ANZSCPB): DM 78.-

Preface

Though it is a pleasure to write a short foreword to this collection of excellent scientific papers covering a range of biological topics, the rather depressing feature is the small number of papers. All-electronic publishing is developing and your Editors do have great faith in it. One problem for potential authors has been the reluctance of the abstracting journals to pay any attention to electronic journals – perhaps Springer should make a rapid move in this area and start the first all-electronic journal abstracting this type of literature. However, even the paper citation journals are starting to pay attention to the medium.

The particular advantages of all-electronic publishing are beginning to emerge more clearly and it is clear that publishing video material is a unique advantage of our format.

Several papers took advantage of this – for example those by Riehle and others on cell behaviour in tunnels, by Bereiter-Hahn and Voss on zonation in the plasmalemma and by Pavlikova, Zicha, Chaloupkova and Vesely on cell motility of tumour cells. These papers made essential and extensive use of video material, publishing some material of great originality. The work on cell polarity and calcium ions in *Fucus* embryos by Brownlee, Manison and Anning used animation to present their results in an especially clear way. The facility of use of animation is another special advantage of our type of publication that should be more widely used. It is easy to see that several other areas of biology could make extensive use of EBO and publish studies on animal behaviour and animal mechanics with unique video material.

A still wider range of subjects should be using animation methods to present their results. Video material could even be included in the Materials and Methods sections of many papers to illustrate the particular procedures used. Indeed it should be used if the procedure is novel, difficult or complex. So we invite biologists to take advantage of these special facilities.

We thank our Editorial Boards. During the next few months there will be changes in the composition of those boards as some come to the end of their three year period of service and as new members are recruited.

I and my fellow Editors, Chris Bridges and Dale Sanders, would like to thank the staff of Springer and especially Mr Thomas Bastuck for their work in ensuring that EBO has been published correctly and expeditiously. We also thank our reviewers whose IT skills are kept perfectly honed by the need to read and view the material electronically.

Adam Curtis

Centre for Cell Engineering
Joseph Black Building
Glasgow University
Tel 0141 330 5147
Fax 0141 330 3730

Experimental Biology Online 1998

Table of Contents Vol. 3 (1998)

Effect of fasting and thermal acclimation on metabolism of juvenile axolotls (*Ambystoma mexicanum*)

Louis N. Irwin[1] · Karen A. Talentino[2] · and Denise A. Caruso[3]

[1] Department of Biological Sciences, University of Texas at El Paso, El Paso, TX 79968, USA
[2] Department of Biology, Simmons College, Boston, MA 02115, USA
[3] Department of Immunology, University of Texas, M. D. Anderson Cancer Center, Houston, TX 77030, USA

Received: 7 July 1997 / Accepted: 22 December 1998 / Published: 20 January 1998

Abstract. Juvenile axolotls were acclimated to 15°C or 25°C and either fed or fasted at both temperatures, to study the interaction of thermal acclimation and nutritional state on metabolism. Fasting but not thermal acclimation significantly increased oxygen consumption at 15°C. Fasting also increased the specific activities of two oxidative metabolic enzymes – citrate synthase and cytochrome oxidase – but not that of the glycolytic enzyme lactic dehydrogenase. The specific activity of cytochrome oxidase was further stimulated by cold acclimation. Triglycerides and fatty acids were severely depleted in fasted animals, but thermal acclimation had no significant effect on lipid stores. This study illustrates: (1) the differential nature of various metabolic responses to fasting; and (2) the confounding interaction of the nutritional state on thermal acclimation experiments in an ectotherm.

Key words. *Ambystoma mexicanum* · Axolotls · Citrate synthase · Cytochrome oxidase · Lactic dehydrogenase · Lipids · Metabolic rate · Nutrition · Thermal acclimation

Introduction

The Mexican axolotl, *Ambystoma mexicanum*, is indigenous to the highland lakes of Mexico. Like many amphibian species, its population has significantly dwindled recently (Shaffer, 1989). While human impact probably accounts directly for much of the decline (Gonzalez *et al.*, 1986), the role of natural factors can neither be excluded nor assessed without a detailed understanding of the physiological ecology of the species.

Correspondence to: L.N. Irwin (e-mail: *lirwin@utep.edu*)

The ecosystems inhabited by axolotls occur in generally milder regions of North America, where seasonal fluctuations are not great. The aquatic lifestyle of the species further reduces the thermal range to which it is normally exposed. Yet the species is part of a large genus that ranges from arctic to tropical habitats, and probably has evolved from ancestors subjected to greater thermal ranges (Smith, 1989). The influence of temperature on its energetic characteristics is thus of considerable theoretical interest in the study of the evolution of thermal adaptation. Axolotls have also become a favored organism for many types of animal research, especially in developmental biology, and are widely maintained in the laboratory. Therefore, a better understanding of the energetics of this species during development has significant practical value as well.

Thermal acclimation of metabolic rates in salamanders has been studied extensively, and a variety of patterns have been demonstrated: compensation, inverse compensation, and lack of compensation have all been reported (Rome *et al.*, 1992). More empirical information is required to gain a better understanding of the circumstances under which different adjustments to temperature occur. Nutritional state and metabolism are clearly among the factors that interact with thermal adaptation. Because nutritional state can affect metabolic rate (Feder *et al.*, 1984; Jobling, 1981), animals traditionally have been fasted during studies of thermal acclimation of metabolic rates. Whether the results then reflect acclimation to temperature or to fasting is often uncertain. Therefore, the need for controlled experiments in which both thermal acclimation and nutritional state are systematically varied is evident.

In this study, we have carried out an experiment to determine if thermal acclimation alone, nutritional state alone, or an interaction between the two affects metabolism at the organismic level, as measured by resting oxygen consumption. We have further explored the biochemical basis for the observed changes in metabolism at the cellular level by analyzing glycolytic and oxidative enzyme activities, and the mobilization of lipid stores. To maximize the utility of these findings for the use of axolotls in developmental studies, juvenile animals were used. We hypothesized that acclimation to cold and fasting would affect metabolism at both the organismic and cellular levels. Our results indicated, however, that nutritional state had a much greater impact than cold acclimation on metabolism over the short duration of this acute experiment.

Materials and methods

Animals, nutrition, and metabolic rates

Juvenile axolotls weighing 4.4 $\pm$ 1.2 g (mean $\pm$ SD) were obtained from Carolina Biological Supply in November. Each animal's oxygen consumption was measured at 20°C to determine empirically the relationship between baseline oxygen consumption and body weight. Animals were individually removed from their home tanks and placed in a closed vessel for measurement of oxygen consumption. The dissolved oxygen content of the water was measured at the start and at the end of a 15-min period with an oxygen electrode coupled to a Cameron Dual-Channel oxygen meter. Animals were maintained at rest throughout the measurements, which were made near midday when they normally showed some activity; thus our measurement of oxygen consumption corresponds most closely to an estimate of resting metabolic rate as defined by Gatten *et al.* (1992). The data were fit to a non-linear regression equation by GraphPad Prism 2.0 to determine the exponent of body weight predictive of oxygen consumption.

Each animal was then randomly assigned to an acclimation temperature (AT) of 15°C or 25°C for 3 weeks, based on preliminary data showing this to be the minimal time required for an apparent difference in oxygen consumption to develop in response to different diets. A duration as short as possible was chosen to minimize the effects of food deprivation. Half the axolotls (n=9) at each AT were fed individually twice weekly with three food pellets (six pellets per week), while the other half were fasted. All nine animals survived in each of the two cold acclimation groups, while eight survived in the fasted group acclimated to 25°C, and six survived in the fed group acclimated at 25°C. At the end of the acclimation period, oxygen consumption was measured for each animal at test temperatures of 15, 20, and 25°C, after which it was killed by decapitation. Two sets of tissue samples from liver and tail muscle were collected from each animal and either processed for enzyme analysis or frozen for later lipid extraction.

Enzyme activity

Fresh tissues were weighed and transferred to a Potter-Elvehjem homogenization tube to which 2.0 ml ice-cold 0.25 M sucrose in 100 mM potassium-phosphate buffer (KPB) was added. For the first enzyme assay, that of cytochrome oxidase (COX), the pH was adjusted to pH 6.0 and a fresh preparation was used. Tissues were homogenized for 1–2 min to a uniform consistency, transferred to a plastic centrifuge tube, and spun at 1000 *g* for 10 min to pellet a crude nuclear fraction which was discarded. Aliquots of the supernate were taken for analysis of COX activity, or frozen for subsequent analysis of citrate synthase (CS) and lactic dehydrogenase (LDH) activity. Assays were determined by the method of Kleckner and Sidell (1985) for measuring maximal enzyme activities, using a Beckman single-beam spectrophotometer.

The specific activity of LDH was determined at 21°C by adding 50-ml aliquots of supernate containing enzyme to 0.65 ml imidazole buffer (76 mM, adjusted to pH 7.6) and 0.1 ml of 3.3 mM pyruvate. The reaction was started by adding 0.2 ml of 0.75 mM NADH. Change in absorbance at 340 nm was measured over a 60-s period.

The specific activity of CS was determined on triplicate 50-ml aliquots of supernatant from the tissue homogenate added to 0.65 ml imidazole buffer (76 mM, adjusted to pH 7.6 at 15°C), 0.1 ml of 5,5'-dithiobis[2-nitrobenzoic acid] (DTNB, 1 mM), and 0.1 ml oxaloacetic acid (OAA, 2.5 mM) or water in alternate tubes. The reaction was started by adding 0.1 ml acetyl CoA (1.5 mM). Enzyme activity was based on the difference in change in absorbance with and without the addition of OAA over a 25-min period at 21°C.

The specific activity of COX in the supernatant from fresh tissue homogenate was measured. To a mixture consisting of 0.85 ml KPB (100 mM, pH 6.0) and 50 ml aliquot of supernate, was added 0.1 ml of 0.25 mM cytochrome-c in KPB (10 mM, pH 7.0) containing 1 mM ethylenediaminetetraacetic acid (EDTA), reduced by sodium dithionite and prepared by gel sieving with Sephadex G-25 (Helmerhorst and Stokes, 1980), to start the reaction. Enzyme activity was calculated from the change in absorbance at 550 nm over 20 min at 21°C.

Lipid analysis

Frozen tissue samples whose wet weights were 0.14 g or less were homogenized in 1.0 ml chloroform:methanol (2:1, v:v) for at least 1 min in a motor-driven Potter-Elvejehm apparatus for lipid extraction as described by Irwin and Schwarz (1983). A 1-cm column of dry silica gel G in a Pasteur pipet plugged with glass wool was pre-rinsed with chloroform:methanol (2:1). The total tissue homogenate was transferred to the silica gel column, then the eluate was collected and reapplied. The homogenization and collection tubes were both rinsed with chloroform:methanol (2:1, v:v) and added to the column to give a total elution volume of ≈ 2 ml. The eluate containing neutral and some acidic lipids was collected again. Aliquots corresponding to 1.5 mg liver or 6.0 mg muscle were dried down, resuspended in 20 μl chloroform:methanol (2:1, v:v), and spotted in 4-mm lanes 2 cm from the bottom edge of 10 cm × 20 cm glass plates coated with 0.25 mm silica gel 60 (EM Science). The plates were developed for 15 min in hexane:ethyl ether:glacial acetic acid (70:30:1, v:v:v), dried, sprayed with concentrated sulfuric acid: water (1:1, v:v), and heated to 130°C for 30 min. Spots corresponding to cholesterol, free fatty acids, triglycerides, and cholesterol esters were identified by co-migration with standards, and quantified by video densitometry (Biomed Instruments, Fullerton, Calif., USA). The densitometric intensity of each lipid band on the chromatography plate reflects the relative concentration of that lipid in proportion to the other lipids in the sample, and allows comparison of each lipid between the fed and fasted groups.

Statistical analysis

These experiments were designed to test the effects of both thermal acclimation and nutritional state, and their possible interaction. Thus data were first subjected to a two-way analysis of variance (ANOVA). If either independent variable was found not to have a significant effect, data for that variable were pooled and tested for a significant influence of the other variable by a *t*-test, or, if variances were found to be unequal, by the Mann-Whitney *U*-test.

Results

Effect of thermal acclimation and fasting on body mass and metabolic rate

Over the 3 weeks of the experiment, fed animals gained weight, while fasted animals lost weight (Fig. 1). To compensate for possible changes in oxygen consumption attributable to changes in body mass alone, standard metabolic rates were established at the start of the experiment to determine empirically the relationship between oxygen consumption and body weight, since the relationship is known to be non-linear (Gatten *et al.*, 1992). Oxygen consumption was found to vary with body weight to the power 0.83 (Fig. 2) – a value very close to that typically found for urodeles (Gatten *et al.*, 1992). Subsequent calculations of resting oxygen consumption thus were corrected by weight in $g^{0.83}$.

The effects of both thermal acclimation and nutritional state on metabolic rate were measured at test temperatures corresponding to the two acclimation temperatures (15°C and 25°C) and an intermediate temperature of 20°C. The results at the lowest and highest test temperatures are shown in Fig. 3. (Results at 20°C were comparable to those at 25°C.) At the lower test temperature of 15°C, fasting resulted in a significant increase in resting oxygen consumption at both acclimation temperatures ($F = 40.7$, $df = 1/22$, $P < 0.0001$). At the higher test temperature of 25°C, neither nutritional state nor thermal acclimation resulted in different values of resting oxygen consumption.

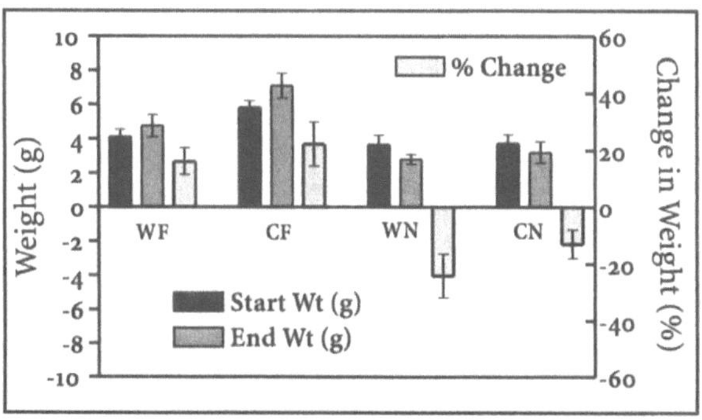

Fig. 1 Body weights of axolotls (g, *left axis*) at start (■) and end (▭) of experiment on warm-acclimated, fed (*WF*), cold-acclimated, fed (*CF*), warm-acclimated, non-fed (*WN*) or cold-acclimated, non-fed (*CN*) animals. The change in body weight (% of starting weight, *right axis*) is shown (▭) for each group over the course of the experiment

Fig. 2 Oxygen consumption as a function of body weight, determined for all animals ($n = 32$) at 20C at the start of the experiment. The *line* shows the best fit for the data to the theoretical relationship between oxygen consumption (y) and body weight (x) of $y = ax^b$

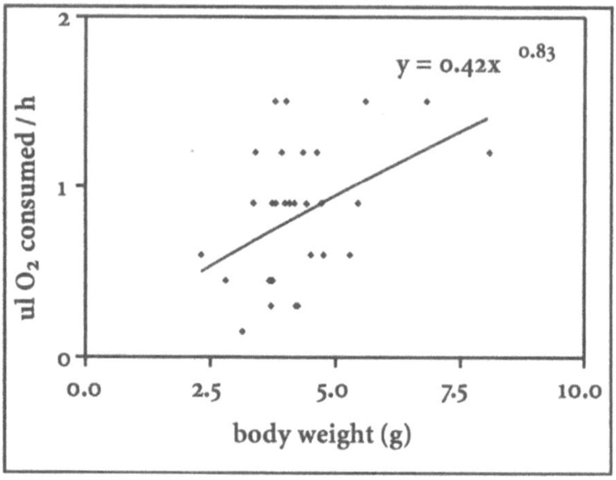

Fig. 3. Effect of thermal acclimation and food intake on oxygen consumption. Axolotls were fed or fasted for 3 weeks during exposure to 25 or 15°C. Oxygen consumption was measured at the end of the experiment at 15 and 25°C. *Bars* represent means ± SEM for the number of animals in each group indicated across the top of the graph. (Sample sizes are less than number of survivors due to rejection of a few faulty baseline oxygen readings)Effect of thermal acclimation and fasting on metabolic enzyme activity

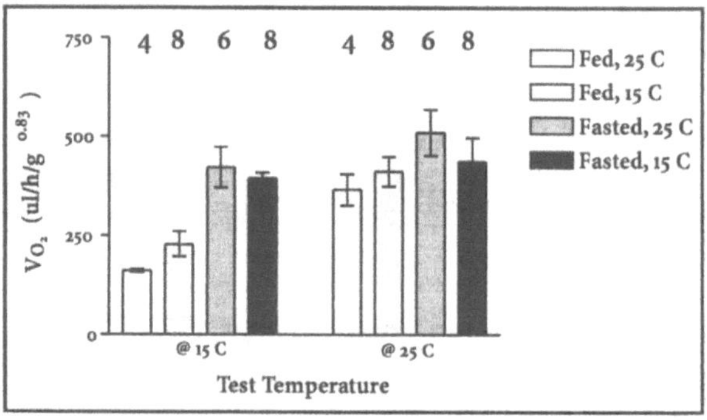

Fig. 4. Effect of nutritional state on lactic dehydrogenase (*LDH*) and citrate synthase (*CS*) in liver and muscle of fed and fasted axolotls. The probability of differences in specific activity of CS between fed and fasted groups was determined by two tailed *U*-tests

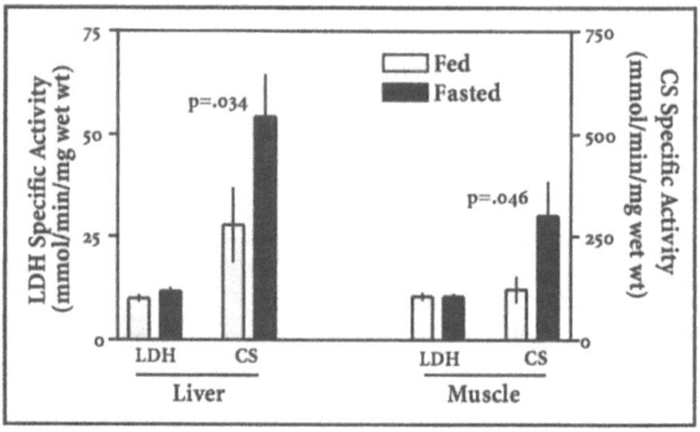

Effect of thermal acclimation and fasting on metabolic enzyme activity

The specific activities of all three enzymes were first analyzed by two-way ANOVA to test for the effect of both nutritional state and cold acclimation. In neither liver nor muscle were LDH or CS affected by thermal acclimation; therefore, the data from cold- and warm-acclimated groups for these two enzymes were combined to test the effect of fasting alone (Fig. 4). Since variances were unequal between the pairs of data sets compared, significance was tested with the non-parametric *U*-test. That analysis shows that fasting had no effect on LDH, but did result in elevated CS activity in both liver ($U = 40$, $n_1 = 10$, $n_2 = 14$, $P = 0.034$) and muscle ($U = 34$, $n_1 = 10$, $n_2 = 12$, $P = .046$).

Because ANOVA revealed significant effects of both fasting and cold acclimation on the activity of COX, the specific activities for this enzyme were plotted individually for the four treatment groups in muscle and liver (Fig. 5). The interaction between nutritional state and thermal acclimation is obvious for both liver ($F = 14.99$, $df = 1/28$, $P = 0.0006$) and muscle ($F = 16.88$, $df = 1/28$, $P = 0.0003$). In both tissues, significant increases in COX activity were restricted to the cold-acclimated, fasted animals.

Effect of thermal acclimation and fasting on lipid stores

ANOVA revealed a significant effect of the nutritional state on lipid stores, but no effect of thermal acclimation and no interaction, in either liver or muscle. Therefore, data for the two thermal acclimation groups relating to all lipid fractions were pooled for comparing the effect of nutritional state. In liver (Fig. 6), fasting had no effect on cholesterol, but resulted in significant depletion of fatty acids ($P<0.003$), and triglycerides ($P<0.001$).

Also in muscle (Fig. 7), fatty acids were significantly ($P<0.05$) depleted in fasted animals. Mean differences in triglyceride concentrations in the same direction were not significant ($P<0.08$) because of a large variance in the data from the fed group.

Discussion

The mean resting oxygen consumption in resting juvenile axolotls ranged from 0.092 to 0.178 ml O_2 g^{-1} h^{-1}, depending on experimental treatment. These values are similar to that reported for *A. mexicanum* by Lenfant *et al.*, (1970), and higher than that reported by Gahlenbeck and Bartels (1970). They fall within the range reported for a number of different ambystomids (Burggren and Wood, 1981; Heath, 1976; Whitford and Hutchinson, 1965, 1967).

Fasted animals had higher resting oxygen consumption values than those that were fed (Fig. 3) when tested at 15°C but not at 25°C. Some authors have suggested that starvation lowers metabolic rate (Merkle and Hanke, 1988),

Fig. 5. Effect of thermal acclimation and nutritional state on cytochrome oxidase (COX) activity. *Bars* represent means ± SEM for $n = 6–9$

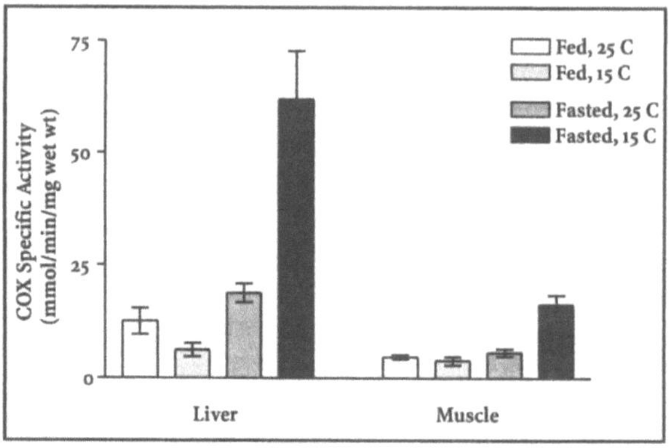

Fig. 6. Relative concentration of cholesterol (*CH*), free fatty acids (*FA*), triglycerides (*TR*), and cholesterol esters (*CE*) in liver tissue of fed and fasted axolotls. *Bars* represent means + SEM in optical density units for $n = 10$

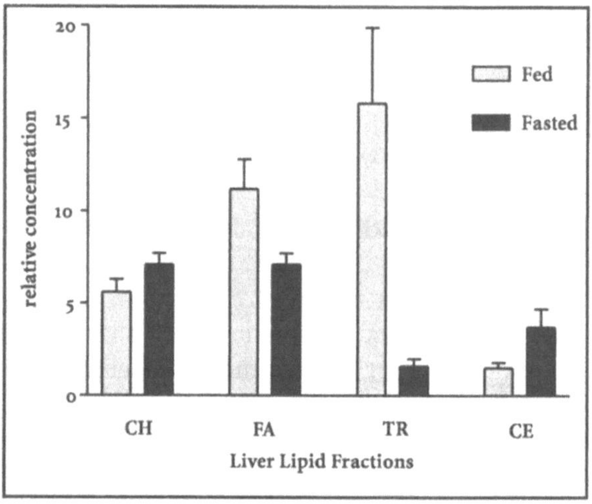

Fig. 7. Relative concentration of cholesterol (*CH*), free fatty acids (*FA*), triglycerides (*TR*), and cholesterol esters (*CE*) in muscle tissue of fed and fasted axolotls. *Bars* represent means + SEM in optical density units for $n = 10$

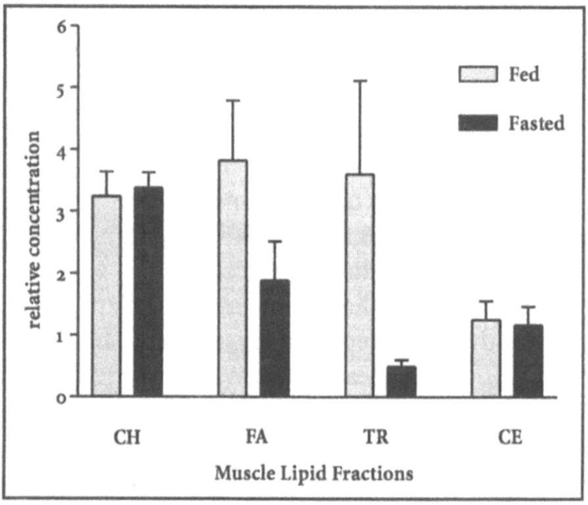

while others have claimed that nutritional state has no effect (Fromm and Johnson, 1955). Our results suggest that fasting increases oxidative metabolism in axolotls at lower ambient temperatures. The absence of this difference at a higher test temperature (25°C) may explain the different results reported in the literature. Obviously care must be taken to use comparable test temperatures when comparing results across studies.

The three enzyme activities measured in this experiment had three patterns of response. The activity of LDH, a glycolytic enzyme, did not respond to either cold acclimation or fasting. The specific activities of both CS (Fig. 4), an enzyme of the oxidative citric acid cycle, and COX (Fig. 5), an enzyme of the oxidative electron transport chain, were increased by fasting; though for COX but not CS, the activity was elevated specifically in cold-acclimated, fasted animals. The lack of a thermal acclimation effect on LDH confirms similar observations made of other amphibians (De Costa *et al.*, 1981; Feder, 1983; Tsugawa, 1980). While LDH activity is not a quantitative correlate of glycolysis, the absence of a fasting effect on the specific activity of this enzyme is consistent with the view that fasting does not involve changes in glycolytic metabolism. On the other hand, the increase in specific activity of CS and of COX in fasted animals is consistent with the increased resting oxygen consumption values associated with chronic fasting (Fig. 3). Under the conditions of this experiment, it appears that the axolotls compensated for the lack of food intake by shifting to more efficient oxidative metabolism.

The reason for the differential effect of cold acclimation on the two oxidative enzymes is unclear. COX but not CS appears to be particularly sensitive to cold acclimation under fasting conditions. Perhaps the 3-week duration of the experiment was too brief for a similar difference in CS to develop. Obviously, the effect of diet on enzyme activities is more temperature sensitive for some enzymes than for others. These results emphasize the fact that the activity of no single enzyme can be taken as a generic metabolic response to either diet or thermal acclimation.

The dramatic depletion of fatty acids and triglycerides in the fasted animals (Figs. 6 and 7) attests to the importance of these lipids for maintenance of metabolism when caloric intake is minimized. Much of the increase in resting oxygen consumption thus may be a reflection of lipid oxidation for maintenance. The lack of change in cholesterol presumably reflects the contribution of cholesterol to stable membrane structures. More subtle changes in lipid composition demonstrated in other studies of amphibians exposed to chronic cold (Feder and Gibbs, 1982; Irwin and Talentino, 1990) probably reflect physicochemical adjustments at the cellular level unrelated to maintenance metabolism.

Caution should be exercised when drawing conclusions from this study. To minimize the detrimental effects of fasting, the experimental period was restricted to only 3 weeks. This may have been insufficient time for the effects of thermal acclimation to develop. The study also focused on biochemical responses at the cellular level, so changes in the liver to body mass ratio (hepatosomatic index) were not considered. These changes may well affect the response of the whole animal in a way not reflected at the cellular level (Kent *et al.*, 1988). Thus, while our results demonstrate that changes take place in enzyme activities and lipid stores per unit of liver or muscle mass, irrespective of the response of the whole organ, a full understanding of the integrative response of the intact animal to thermal acclimation and nutritional state would have to take into account changes in total organ mass in relation to whole body mass. Finally, seasonal variations were not considered in this study. Axolotls in nature do not experience significant seasonal variations in temperature, and the animals used in our study were laboratory bred from a commercial supplier, so seasonal fluctuations are less likely to be a confounding variable than in a captured wild temperate zone species such as the channel catfish, in which seasonal fluctuations have been shown to be a significant factor in acclimation (Seddon and Prosser, 1997).

Attention has been drawn by others to the confounding effects of the nutritional state in thermal acclimation experiments (Baranska and Wlodawer, 1969; Feder *et al.*, 1984). Because of differences in the rate of digestion and the consequent differences in rates of energy metabolism at different temperatures (Feder *et al.*, 1984), thermal acclimation experiments on amphibians have traditionally been conducted using fasting animals. Our results suggest that fasting itself has a more dramatic effect on metabolism than does thermal acclimation. And when the effect of thermal acclimation *is* observed, as in the activity of COX (Fig. 5), it is exacerbated by fasting. Our data illustrate how the apparent effect of thermal acclimation may well be an artifact of nutritional state, as cautioned by Feder *et al.* (1984).

Acknowledgements. This work was supported by NSF grants BNS-8819801 and BNS-9046692 to L.N.I. Advice on quantitative analysis was provided by the Statistical Consulting Laboratory at the University of Texas at El Paso, supported by RCMI grant 2G12-RR08124 from the National Center for Research Resources, NIH. We thank Paula Marcoux, Giselle LeBlanc, Kathleen Bouffard, Jackie Cohen, and Michele Casoli for help with early phases of the work.

References

Baranska, J., Wlodawer, P. (1969) Influence of temperature on the composition of fatty acids and on lipogenesis in frog tissues. *Comp Biochem Physiol* **28:** 553–570

Burggren, W.W., Wood, S.C. (1981) Respiration and acid-base balance in the salamander *A. tigrinum*. Influence of temperature acclimation and metamorphosis. *J Comp Physiol* **144:** 241–246

De Costa, J., Alonso-Bedate, M., Fraile, A. (1981) Temperature acclimation in amphibians: changes in lactate dehydrogenase activities and isozyme patterns in several tissues from adult *Discoglossus pictus pictus*. *Comp Biochem Physiol* **70B:** 331–339

Feder, M.E. (1983) Metabolic and biochemical correlates of thermal acclimation in the rough-skinned newt *Taricha granulosa*. *Physiol Zool* **56:** 513–522

Feder, M.E., Gibbs, A.G. (1982) Thermal acclimation of oxygen consumption, enzyme kinetics, and lipids in salamanders. *Am Zool* **21:** 962

Feder, M.E., Gibbs, A.G., Griffith, G.A., Tsuji, J. (1984) Thermal acclimation of metabolism in salamanders: fact or artifact. *J Therm Biol* **9:** 255–260

Fromm, P.O., Johnson R.E. (1955) The respiratory metabolism of frogs as related to season. *J Cell Comp Physiol* **45:** 343–359

Gahlenbeck, H., Bartels, H. (1970) Blood gas transport properties in gill and lung forms of the axolotl (*Ambystoma mexicanum*). *Respir Physiol* **9:** 175–182

Gatten, R.E. Jr., Miller, K., Full, R.J. (1992) Energetics at rest and during locomotion. In: Feder, M.E., Burggren, W.W. (eds) *Environmental physiology of the amphibians*. University of Chicago Press, Chicago, pp. 314–377

Gonzalez, A., Arturo, H., Camarillo, J.L., Mendoza, R.F., Mancilla, M.(1986) Impact of expanding human populations on the herpetofauna of the Valley of Mexico. *Herpetol Rev* **17:** 30–31

Heath, A.G. (1976) Respiratory responses to hypoxia by *Ambystoma tigrinum* larvae, paedomorphs, and metamorphosed adults. *Comp Biochem Physiol* **55A:** 45–49

Helmerhorst, E., Stokes, G.B. (1980) Microcentrifuge desalting: a rapid quantitative method for desalting small amount of protein. *Anal Biochem* **104:** 130–135

Irwin, L.N., Schwartz, K. (1983) Amphibian brain gangliosides: pattern analysis by two dimensional thin layer chromatography. *Comp Biochem Physiol* **76B:** 649–651

Irwin, L.N., Talentino, K.A. (1990) Brain ganglioside patterns and effect of thermal acclimation in North American salamanders. *Comp Biochem Physiol* **96B:** 471–473

Jobling, M. (1981) The influences of feeding on the metabolic rate of fishes: a short review. *J Fish Biol* **16:** 629–638

Kent, J., Koban, M., Prosser, C.L. (1988) Cold-acclimation-induced protein hypertrophy in channel catfish and green sunfish. *J Comp Physiol* **158B:** 185–198

Kleckner, N.W., Sidell, B.D. (1985) Comparison of maximal activities of enzymes from tissues of thermally acclimated and naturally acclimatized chain pickerel (*Esox niger*). *Physiol Zool* **58:** 18–28

Lenfant, C., Johansen, K., Hanson, D. (1970) Bimodal gas exchange and ventilation-perfusion relationship in lower vertebrates. *Fed Proc* **29:** 1124–1129

Merkle, S., Hanke, W. (1988) Long-term starvation in *Xenopus laevis* Daudin: 1. Effects of general metabolism. *Comp Biochem Physiol* **89B:** 719–739
Rome, L.C., Stevens, E.D., John-Alder, H.B. (1992) The influence of temperature and thermal acclimation on physiological function. In: Feder, M.E., Burggren, W.W. (eds) *Environmental physiology of the amphibians.* University of Chicago Press, Chicago, pp. 183–205
Seddon, W.L., Prosser, C.L. (1997) Seasonal variations in the temperature acclimation response of the channel catfish, *Ictalurus punctatus. Physiol Zool* **70:** 33–44
Shaffer, B.H. (1989) Natural history, ecology, and evolution of the Mexican „axolotls". *Axolotl Newslett* **18:** 5–11
Smith, H.M. (1989) The axolotl in its native habitat. *Axolotl Newslett* **18:** 12–16
Tsugawa, K. (1980) Thermal dependence in kinetic properties of lactate dehydrogenase from the African clawed toad, *Xenopus laevis. Comp Biochem Physiol* **66B:** 459–466
Whitford, W.G., Hutchinson, V.H. (1965) Gas exchange in salamanders. *Physiol Zool* **38:** 228–242
Whitford, W.G., Hutchison, V.H. (1967) Body size and metabolic rate in salamanders. *Physiol Zool* **40:** 127–133

Cell behaviour in tubes

Mathis Riehle · Diana Ferris · Douglas Hamilton · Adam Curtis

Centre for Cell Engineering, University of Glasgow, Glasgow G12 8QQ, UK

Received: 14 October 1997 / Accepted: 26 January 1998 / Published: 6 February 1998

Abstract. The motile behaviour, growth and reactions of osteoblasts, chondrocytes, epitenal cells and endothelia to culture on extended concave surfaces inside tubes is reported. Time-lapse video observations show that these cells can be successfully grown in quartz tubes for up to 21 days. The cells would attach to the concave surfaces even when the radius of curvature was as low as 25 μm. The cells attached to the walls were not oriented. As culturing progressed the chondrocytes, osteoblasts and epitenon cells tended to detach from the walls of the tubes, forming long cords of cells attached to the walls at a few points only. Endothelial cells did not detach. These cords showed extensive lamellipodial activity at their points of attachment to the tube wall. The cells in these cords do not show contact inhibition of movement.

Key words. Bone · Cell behaviour · Chondrocyte · Endothelia · Epitenon

Introduction

Various cell types in animal tissues live in tubes; for example, endothelia in blood vessels, synovial and epitenal cells in tendons and bone cells in osteons. In bone the osteoblasts form successive layers of aligned mineral and collagen in concentric circles with the youngest layer as the innermost one. After a period the tube is narrowed to the point at which further mineral deposition ceases – is control of this related to tube diameter and to the concavity of the surface? In endothelia the tubes enlarge by insertion of the enlarging cells and their daughters. Chondrocytes live in an avascular environment of which they are almost the sole architect. These cells are encapsulated in columns termed chondrons, which are surrounded by extracellular matrix. This is a constant, balanced environment for the chondrocyte which is relatively inaccessible to metabolites, the cells gaining nutrients only by diffusion (Von der Mark, 1986).

Correspondence to: A. Curtis (email: A.Curtis@biol.gla.ac.uk)

There are many reports (Selye *et al.*, 1960; Bobyn *et al.*, 1980; Eggli *et al.*, 1997) on the reactions of cells to square section or V-section grooves, summarised by Chehroudi and Brunette (1995) and by Curtis and Wilkinson (1998), and to convex surfaces (e.g. Curtis and Varde, 1964; Dunn and Heath, 1976) but few observations on the reactions of cells to concave surfaces have been made. There have been few studies of cell behaviour in tubes though the early work of Twitty and Niu (1954) and Dunn (1982) is noteworthy.

The study of osteoblasts in tubes could provide information hitherto unattainable about the optimal implant coating dimensions for the orthopaedics industry. Work on the practical aspects of achieving good osseo-integration with metallic, ceramic or polymeric surfaces in prostheses has led to the suggestion that effective pore sizes for encouraging bone growth are in the range of 40 µm to 600 µm (Brunette, 1988; Gray *et al.*, 1996). However, the opaque materials used in these studies on prostheses do not permit continuous observation of osteoblast behaviour and mineralisation, nor can they examine the effect of a single pore size on the parameters, whilst the quartz tube system provides a precise topography and can be monitored continuously during culture.

We report on the behaviour of epitenal, endothelial, chondrocytic and osteogenic cells in transparent quartz tubes using time-lapse video techniques to describe and analyse their behaviour and reaction to concave surfaces. It is hoped to publish other reports on mineralisation and cytoskeletal organisation elsewhere, though light-microscope observations relevant to mineralisation will be described here.

Materials and methods

Capillary tubes for culture

The thin tubes (40–300 µm inner diameter) were drawn manually by W. McCormack (Glassblowing Workshop of the University of Glasgow), to whom we are much indebted, from fused quartz material and were optically isotropic. The wide tubes (inner diameter >700 µm), X-ray diffraction tubes (Müller, Berlin, Germany) were used without modification. The tubes were regarded as being cleaned by the high temperature of fabrication and were sterilised by infusion of 70% ethanol followed by replacement of the ethanol

Table 1. Tube dimensions

Tubes	Inner diameter	Length
Group A	150–300 µm	2–5 mm
Group B	700 µm	0.5–2.3 cm

with distilled sterile water which in turn was allowed to evaporate under sterile conditions before filling the tubes with cells. All 700 μm tubes (Group B) were a gift from Neil Issacs (W. Mueller, Germany) (Table 1).

Cell culture

Osteoblasts. Isolation of calvaria from neonate Sprague-Dawley rat pups was performed aseptically using a modified version of the protocol described by Hung *et al.* (1995). The endosteum and periosteum were scraped from the frontal, parietal and occipital bones, which were chopped into 1- to 2-mm-long fragments, and were vortexed in Dulbeccoís modified Eagleís digestion medium (DMEM, supplemented with 6 mM glutamine, 200 U/ml penicillin, 200 μg/ml streptomycin, 5 μg/ml amphotericin B) until fragments appeared white. Five 20-min periods of digestion at 37 °C using collagenase (2 mg/ml, Type VIII or Type IA, Sigma, Poole, Dorset, UK) in DMEM digestion medium followed. The first cell population released was discarded. After each digestion, the supernatant was collected, culture medium (DMEM supplemented with 10% calf serum and 3 mM glutamine, 100 U/ml penicillin, 100 μg/ml streptomycin, 2.5 mg/ml amphotericin B) was added, and the suspension centrifuged. The sedimented cells were resuspended in culture medium and plated into 75 cm^2 culture flasks. Media was changed on day 1 after plating and every 48–72 h afterwards; at each media change fresh ascorbic acid (0.3 mM) was added. Either primary cells or cells no greater than passage III were used for experiments. Neither ascorbic acid nor β-glycerophosphate was added to the medium once the cells were placed in the tubes except for the results reported in Table 2, where medium with 10 mM β-glycerophosphate and 0.3 mM ascorbic acid was added on day 7 of culture and every 2 days thereafter.

Chondrocytes. Isolation of chondrocytes from ovine metacarpal-phalangeal joints was performed aseptically according to the standard operating procedure of Smith and Nephew (York Science Park, York, UK). Cartilage was removed from the metacarpal-phalangeal joints and stored overnight at 37 °C in culture medium (DMEM supplemented with 10% fetal calf serum, 3 mM glutamine, 100 U/ml penicillin, 100 μg/ml streptomycin, 2.5 mg/ml amphotericin B and 67 U/ml of nystatin). The tissue was chopped into 1- to 4-mm-long fragments and washed three times with culture medium. The medium was removed and the cartilage was gently shaken in a solution of 1% Pronase (Sigma) in culture media for 1 h at 37 The Pronase solution was aspirated and any cells released at this stage were discarded. The cartilage was then placed in a solution of crude collagenase 300 U/ml (type 1A, Sigma) in culture medium and gently shaken for 3 h at 37 °C until the cartilage was completely digested. The resulting cell suspension was passed through a 70-μm pore nylon cell strainer (Falcon, Becton Dickinson UK, Cowley, Oxford, UK). The filtrate was then transferred into centrifuge tubes and spun at 173 *g* for 5 min. The resulting pellet was washed 3 times and then resuspended in culture medium. The cells were plated in 75 cm^2 culture flasks at a density of 50×10^3 cells/ml and 50 μg/ml ascorbic acid added to the medium. The cells were split fed every 3 days with 50% fresh medium until they reached confluence. Cells were used for experiments up to the second passage.

Endothelia. Cells of the human HGTFN capillary endothelial line isolated in this laboratory. This line is characterised by low-density lipoprotein (LDL) binding, factor VIII production, high levels of angiotensin-converting enzyme and *Ulex europaeus* lectin binding as well as „cobblestone" morphology at confluency.

Epitena. Rat epitenon cells were cultured by the methods described by Wojciak *et al.* (1995).

Culture in capillary tubes. The four types of cell were isolated from culture as single-cell suspensions by trypsinisation and resuspended in serum-containing medium. They were seeded at high density (approximately 3×10^6 cells/ml) by capillary action into two types of quartz tubes (Table 1). Cells in tubes and on control flat surfaces of fused silica were videotaped for extended periods. Cells were fixed and processed for fluorescence and polarised light microscopy techniques for various products (see below).

Video time-lapse phase-contrast micrography

Cell types were videotaped just after seeding for periods of up to 12 days at 37 °C. Videos were recorded with 20× phase-contrast objective magnification using a COHU 4905 CCD camera or a Panasonic Neuvicon video camera under the lowest light level that would produce a good image. Images were recorded on a Panasonic Model AG-6730 time-lapse video recorder at one frame per minute.

Polarisation microscopy

This was carried out using a Leitz polarisation microscope with strain-free optics and a 1λ plate when appropriate.

Analysis

All cell videos were imported into a Macintosh Power PC 7500/100 and analysed using the public domain NIH Image program (developed at the U.S. National Institutes of Health and available on the Internet at: http://rsb.info.nih.gov/nih-image/).

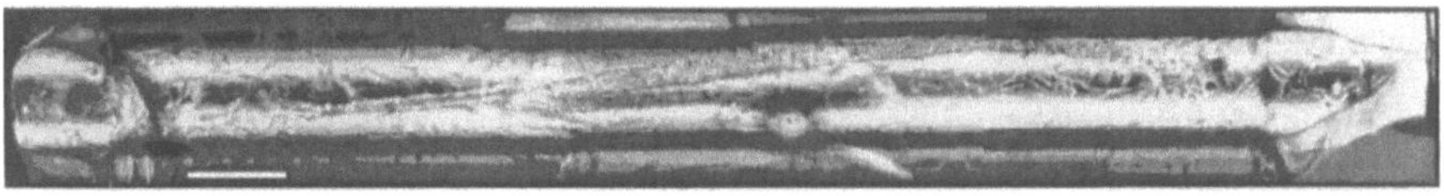

Fig. 1. Epitenon cells in a quartz tube after 3 days. The cells have formed a long cord and individual cells are evident around this cord. *Bar* 200 μm

Immunofluorescence staining of osteocalcin in bone cells

Tubes were rinsed in phosphate-buffered saline (PBS) and fixed in buffered formalin 4% v/v for 1 h. Cells were rinsed in PBS and allowed to remain in PBS at 4 °C overnight. Cells were incubated in a 0.5% PBS solution of bovine albumin (Sigma) for 20 min, stained with anti-human osteocalcin antibody (Biogenesis) 1:25 for 1 h, rinsed 2–3 times with the 0.5% bovine albumin solution and incubated for 1–2 h with FITC-labelled anti-rabbit immunoglobulin (SAPU, Law Hospital Carluke-Lanarkshire, UK) 1:50.

Confocal laser scanning microscopy

Tubes with bone cells in them were examined for the presence of osteocalcin using an Odyssey Real Time Laser Scanning Microscope Model No. VSM-LSM (Noran Instruments, Milton Keynes, UK) equipped with an argon-ion laser, and the MetaMorph Imaging System (Universal Imaging). Settings used were 488 nm excitation, a 0.5 μm aperture, a 15 μm slit, a 40×, NA 0.5, long-working-distance water-immersion objective and a 550 nm barrier filter. Proper controls for second antibody and autofluorescence were tested as well.

Results

General: applying to all cell types used

Viability and division. The endothelia, chondrocytes and osteoblasts showed no evidence of cell death over periods of culture up to 17 days and divisions could be seen in the videos. This was the situation even in the tube centres which were at least 100–200 μm from the mouths of the tubes. Thus, diffusion of nutrients seemed adequate for all of the small-diameter tubes tested. In the very long, i.e. on the order of centimetres, larger diameter (700 μm) tubes there were small regions where cells were not found. In contrast to endothelia, chondrocytes and osteoblasts, most of the epitenal cells moved away from the centres of the tubes fairly rapidly to locate in the mouths of the tubes (i.e. the first 600–1000 μm or so from the ends). A few of these cells remained in situ in the middle and looked dead (Fig. 1).

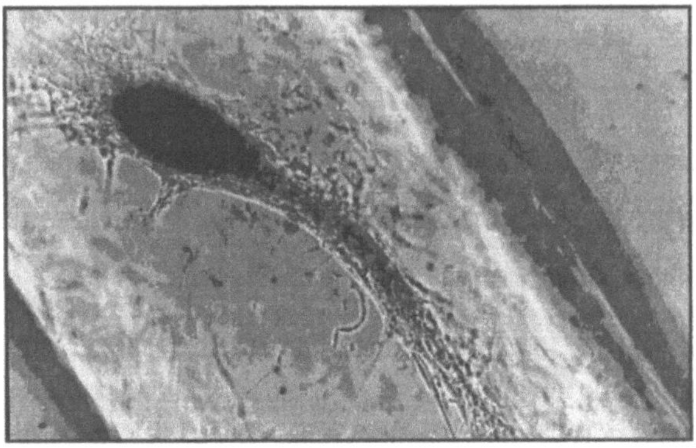

Fig. 2. Osteoblasts fixed after 7 days in culture in 700 μm internal diameter quartz tubes. The bone cells have formed a nodule which is attached to the inner wall of the tube

Movies

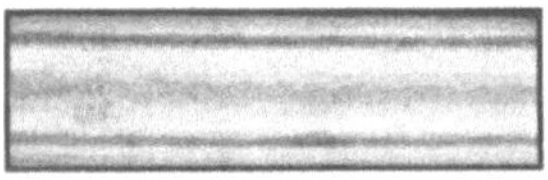

Bone 1. Osteoblasts in 150 μm tubes. The cells can be seen to form nodules which move within the tubes. The osteoblasts also form cords and individual cells can be seen to undergo crab-like movements

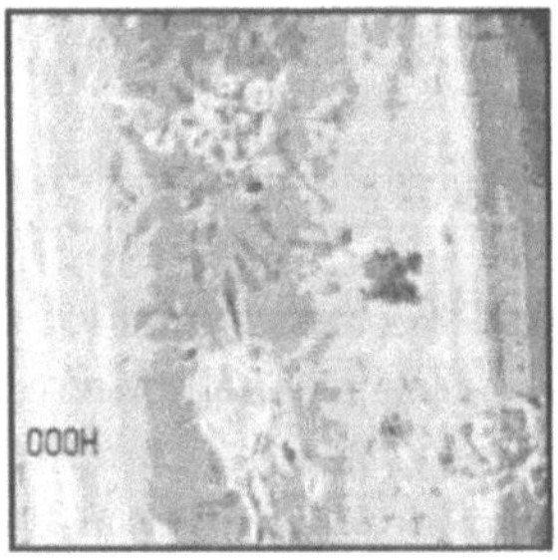

Bone 2. Osteoblasts in 300 μm tubes. Cord formation can be seen as cells join together and these networks move within the tubes

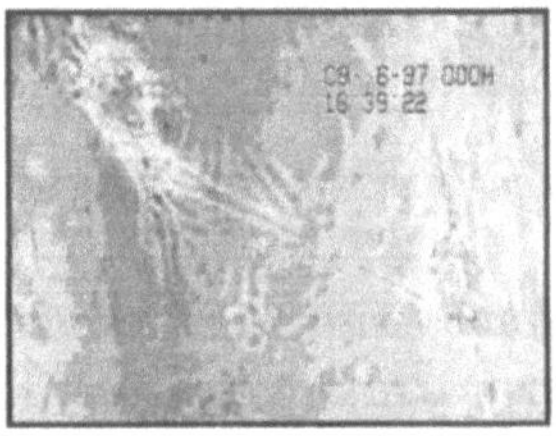

Bone 3. Osteoblasts in 700 μm tubes. The osteoblasts have formed nodules. The two nodules are initially joined, but separate and individual cells can be seen to detach from the nodules. Cells can also be seen to attach and spread, then detach and move

Epitenon Epitenon cells in quartz tubes. The cords formed by the epitenon cells are less dense in comparison with other cell types. The cells detach and reattach to the tubes, but after several divisions there is little movement within the tube

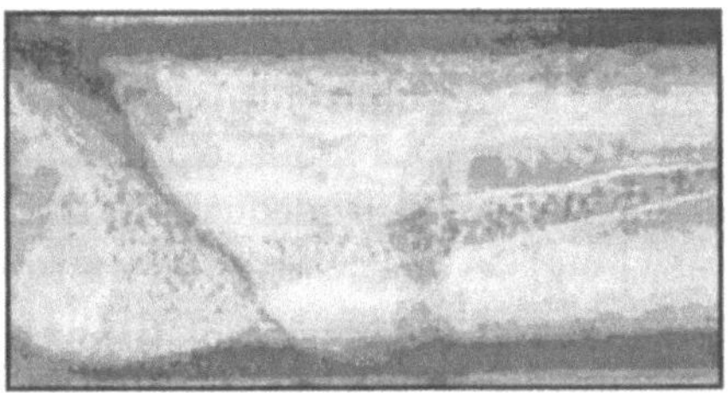

Chondrocyte Ovine chondrocytes in a quartz tube. The cells formed a cord structure which is then seen to disintegrate into single cells, and undergo attachment and reattachment to the tubes. Finally, the chondrocyte cord moves out of the tube, 240 μm internal diameter

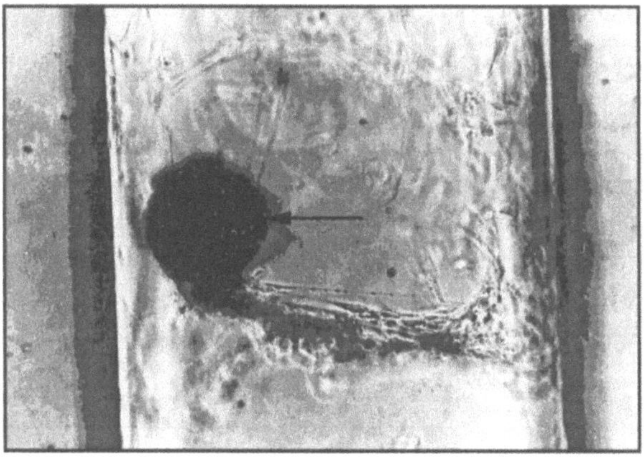

Fig. 3. Osteoblasts fixed after 7 days in culture in a 700 µm internal diameter quartz tube. The cells have formed a nodule which stretches across the entire width of the tube. Individual cells with extended lamellipodia can be seen surrounding this cord of osteoblasts

Morphology and behaviour. The behaviour of epitenon cells, chondrocytes and osteoblasts in tubular structures differs markedly from their behaviour on flat surfaces over the culture in three ways:

In tubes

1. Formation of cords and meshes or networks of cells spanning across the tube. (see Bone Movie 1, Bone Movie 2, Chondrocyte Movie, Fig. 2.)
2. Cell motion ó cells are constantly attaching and reattaching as individuals and as groups of cells (all Movies*).
3. There is no evidence for contact inhibition of movement because the cells in the cord are multilayered.

This differs from the situation on flat surfaces where cells are effectively monolayered and there is little movement once confluence is reached.

Within a few hours of plating, cells stretched out and made contact with one another. The attachments formed span across the tube and along its length. Groups of cells associated into clumps and cords of cells which behaved like a dynamic system, detaching and reattaching at various points along the tube during culture. Nodule-like structures began to appear after the first 24 h of seeding in the osteoblast cultures. These effects can be seen in Bone Movies 1 and 2 and in Figs. 2 and 3.

Cord formation appears to be attained by a series of contractions of groups of cells and an excess of loss over re-establishment of adhesions to the walls. Where the cells contacted the walls very active pseudopodia could be seen in groups of 10–20 connected cells. These very active processes (see Bone Movie 3) give the appearance of spider-like or crab-like motile behaviour. The cords

*All the movies were originally shot at one frame per minute but have been edited down to every 20th frame. The scale can be appreciated from the internal diameters of the tubes quoted in the legends

Table 2. Nodule size and number in single tubes with cells cultured in the presence of β-glycerophosphate and ascorbic acid added at day 5 of culture. Fixed at day 12. Note „*" denotes tube with flared end approximately 1.5 mm wide. All nodules presented were found in 700-mm-wide end. The *nodule number* refers to the number of samples in calculating the standard deviation

Tube length (cm)	Nodule number	Nodule length (μm)	SD	Nodule width (μm)	SD
0.6	1	158	-	74	-
1.2	3	280	57	135	82
1.2	5	198	166	85	24
2.5*	12	124	59	95	31

are appreciably more than one cell thick in each dimension so it can be concluded that contact inhibition of movement is lacking.

In tube cultures of osteoblasts and chondrocytes, viable cells were connected to each other and the cord structures extended over the length of the tube. Epitenal cells formed less densely structured networks (see Epitenon Movie). Movement took place in the cords or networks and this was co-ordinated perhaps by the interplay of mechanical forces. Osteoblast cords were very persistent and were not seen to pull apart. Frequently, chondrocyte cords did pull apart, but were re-formed rapidly (see Chondrocyte Movie). Epitenal cords did not re-form after breaking apart.

The reaction of the cells to the tubular environment seemed to be governed by the diameter and length of the tube. In the smaller diameter tubes, i.e. 150 μm, cells formed cords that attached to each other and ran along the entire length of the tube for as long as 5 mm. However, there is no evidence for any orientation on the tube walls of any of the cell types that we used, unlike the situation on convex surfaces (Curtis and Varde, 1964).

Observations specific to cell type

Osteoblasts. Formation of nodule-like structures (see Figs. 2 and 3) resembling those found in confluent flat cell cultures of osteoblasts took place. These nodules were more frequent per unit length as tube length increased (see Table 2). Linear regression analysis of this small sample showed that the regression was significant at the 2% level. Cells remained viable in equivalent tubes for as long as 30 days and, in some cases, filled the tube with cells and extracellular matrix.

Polarisation microscopy of bone cells

The extracellular matrix production appeared to be enhanced by the tubular environment. Examination under phase-contrast microscopy and polarised

Fig. 4. Osteoblasts in a 280 micron wide and 2.3 cm long tube, fixed after 18 days and stained for osteocalcin. Note the nodules formed along the streak and the highly stained particles along the inner wall of the tube. Confocal laser scanning image at 40×

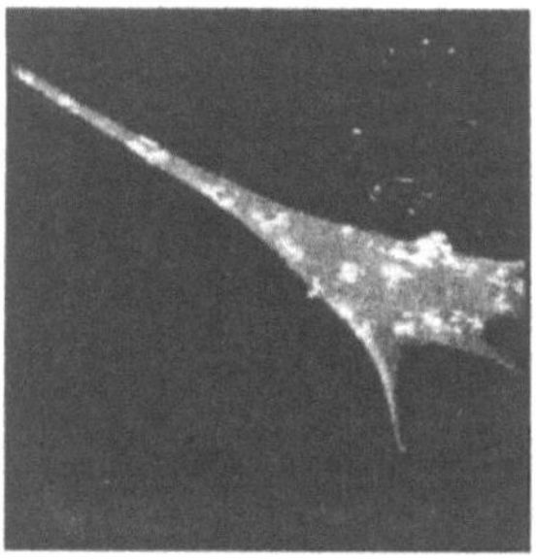

Fig. 5. Osteocalcin staining of an osteoblast cell cord spanning across a tube of 700 μm internal diameter. Cells were fixed at 12 days of culture. Note the small fragments of material along the edge of the cord. Confocal laser scanning image (original magnification 40×)

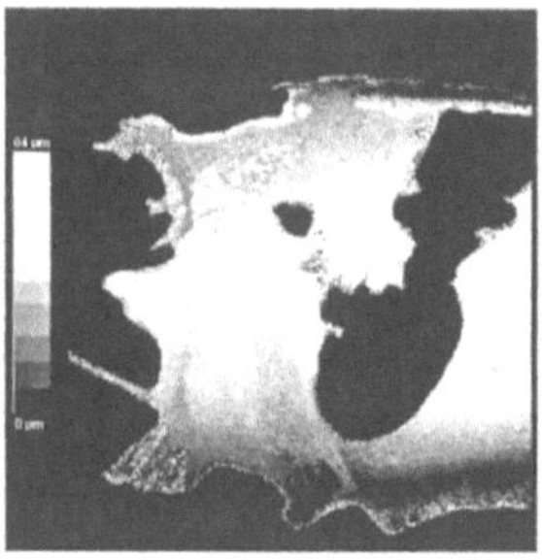

Fig. 6. HGTFN endothelial cells grown in a 280 micron wide quartz tube were fixed after 4 days in culture and stained for F-actin.
The image is a maximum projection of a stack of 65 confocal images covering a total z-distance of 65 μm

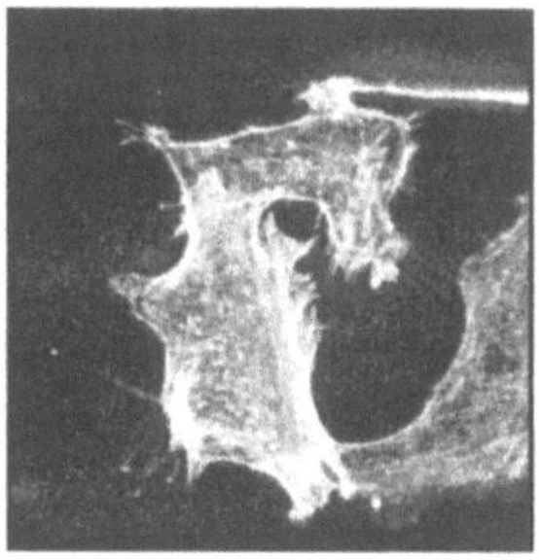

Fig. 7. Derived from Fig. 6. The image is a topographical one derived from the stack of confocal images by finding the brightest pixel at each location, and encoding it with a value that represents the z-position. Therefore, the intensity of the image represents height, with brightest being the highest and darkest the lowest. The image shows that the cells are in the same focus as the inner side of the tube and that these cells spread on the inner side of the tube

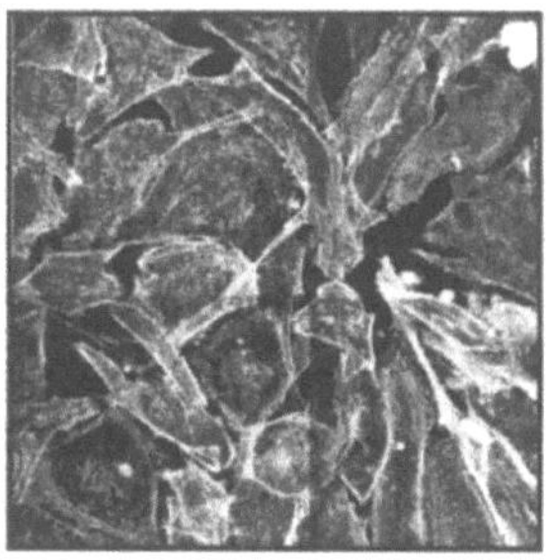

Fig. 8. HGTFN cells on flat quartz surfaces after 4 days of culture. Cells are stained for F-actin. The cells have formed a flat sheet with no obvious actin organisation. Also, less actin appears to be present than was seen with the endothelial cells in the quartz tubes. Same scale as in Fig. 6 and Fig. 7

light revealed that some clumps of cells are similar in morphology to nodules formed *in vitro* (Escarot-Charrier *et al.*, 1988; Aronow *et al.*, 1990). Birefringent particles, most likely a form of hydroxyapatite, were seen in these clumps.

Osteocalcin staining

Osteoblasts in tubes stained positively for osteocalcin (Figs. 4 and 5). The secondary antibody and autofluorescence controls were negative. Small fragments of material were observed along some cords and these stained heavily for osteocalcin, suggesting that this protein might be adsorbed to mineral produced by the cells. In addition, small particles lining the inside of the tube stained brightly for osteocalcin.

Endothelia

Endothelia did not form cords but remained attached to the tube walls and slowly formed very flattened monolayers on the walls, with some degree of cell movement. The cells conformed well to the walls, see Figs. 6 and 7. The cells were more rounded than on flat surfaces (Fig. 8).

Discussion

General observations

Our data show that this type of culture system is not damaging to the epitenal cells, osteoblasts, chondrocytes or endothelia. However, the behaviour of osteoblasts and chondrocytes is changed markedly from that seen on planar culture dishes. It is arguable that the tube provides a culture system much more akin to the natural environment for endothelia, osteoblasts and chondrocytes because the environment is concave in cross-section and the supply of nutrients is restricted as it is naturally for these cells.

It is noteworthy that osteoblasts, chondrocytes or endothelia survive and divide in this form of culture for long periods and appear to maintain their differentiation even though it would be expected that nutrition and removal of waste products would be poor in the tubes. On the other hand, concentrations of autocrine factors would be expected to be higher than would be found in conventional culture systems. No external autocrine factors were added to the cultures, except transferrin for the endothelial cells.

Though accurate determination of cell cycle times in these tube cultures will be difficult it is clear that the cells undergo division, partly because some divisions can be seen in the videos and partly because cell number increases greatly. It should be noted that these remarks do not apply to all cell types because attempts to culture macrophages or fibroblasts in this type of culture system led to rapid cell death.

The movies give the viewer the impression that cell movements are co-ordinated when the cord structures form but this co-ordination may be nothing more than the interplay of mechanical forces balancing out. However, it is noteworthy that in osteoblast cultures mechanical forces acting on the cords and perhaps generated internally rarely break a cord and the same is true to a slightly lesser extent with chondrocytes. The environment appears to be effective for the maintenance or even promotion of the differentiated state for osteoblasts because mineralisation took place under conditions where it would be unlikely to occur in planar culture. The chondrocytes also appeared to maintain their particular morphology well in the tube cultures.

Specific cell types: osteoblasts

The behavioural differences of osteoblasts in tubes versus flat surfaces can be partially explained by the cellular condensation/microenvironment mechanism proposed in several papers by Chehroudi and Brunette (Brunette *et al.*, 1991; Chehroudi *et al.*, 1992; Qu *et al.*, 1996; Chehroudi *et al.*, 1997). These authors performed several studies to look at the influence of pitted surfaces on bone formation. Brunette *et al.* (1991) studied pits with outer surface dimensions of 100 μm × 100 μm with walls tapering to an angle of 55 ° to the surface and an overall depth of 120 μm. Although they were unable to draw definite conclusions, preliminary work suggested that smaller, mineralised nodules appeared more frequently inside the pits. Chehroudi *et al.* (1992) examined larger square pits with 270-μm-long sides. The results of *in vivo* experiments with implants placed in the parietal region of rats revealed mineralisation on some pitted surfaces after 8 weeks whereas no mineralisation was observed next to smooth surfaces. *In vitro* experiments revealed the presence of some large (> 10 μm) nodules on pitted surfaces. Only small (0.5–3 μm) nodules were found on smooth culture surfaces. The authors concluded that the environment created by the topography of their implants positively influenced bone formation both *in vivo* and *in vitro*.

Qu *et al.* (1996) looked at pits with dimensions of 175 μm × 175 μm tapered at 125 ° to 100 μm in depth. The authors suggest that the restricted environment created by the pits may enhance bone formation by establishing a localised region rich in cytokines and extracellular matrix factors. Chehroudi *et al.* (1997) suggest that the geometry of their pits, ranging from 3 to 120 μm in depth and tapered at an angle of 125° to the surface, created an environment of increased cell density with little diffusion of regulatory molecules and limited proliferation. The constraint on growth of cells in high-density environments has been linked to differentiation of various cell types (Boyan *et al.*, 1996). He hypothesises that the curvature of these pits may provide the optimum tension and compression for the cellís mechanoreceptors. Thus, there is already evidence in the literature for the osteoinductive potential of a localised environment such as pits in an otherwise planar surface. We now add evidence that tube culture that naturally resembles the situation in osteons enhances the maintenance and operation of the osteogenic state.

Further support for the creation of a bone-inductive environment by tunnel topography comes from the use of antibodies against osteocalcin. Osteocalcin is a late differentiation marker, appears concomitantly with mineralisation, and has a role in mineral deposition and crystal growth (Gundberg *et al.*, 1984; Robey, 1989; Aubin *et al.*, 1995). When cells in tubes were examined for osteocalcin using confocal scanning laser microscopy, small globules lining the cords and tiny particles lining the tube interior stained heavily. This indicates that the cells in the tubes produce extracellular matrix. Furthermore, given the known binding affinity between hydroxyapatite and osteocalcin, it seems to suggest that the cells are producing small particles of hydroxyapatite.

Conclusions

All cell types responded in slightly different ways to the extended concave surfaces. Epitenal cells lived rather poorly in our tubes and tended to migrate to the ends of the tubes during cell culture, indicating a strong preference for normal cell culture conditions. Endothelia showed a morphology and behaviour similar to that seen in planar cultures.

Osteoblasts and chondrocytes, on the other hand, responded differently to tubular environments when compared to their behaviour on flat surfaces during similar time frames (2–21 days) and to all other cell types tested. Cells formed aggregates including cords that spanned across the tube and nodule-like structures within the osteoblast cord. Constant reattachment and movement of cell cords as a unit were also characteristic behaviour.

The tubular environment appeared to influence extracellular matrix production by osteoblasts in particular. Despite the lack of ascorbic acid and β-glycerophosphate in the majority of our cultures when the cells were in the tubes, components normally necessary for the production of mineralised tissue *in vitro*, osteoblasts appeared to form nodule-like structures similar in morphological appearance to nodules formed by confluent cell layers. Preliminary examination of these structures by polarised light microscopy confirmed the presence of some birefringent particles in these nodules. In addition, osteocalcin staining revealed globules produced along the cells cords and suggested that the inner wall of the tube was coated by small mineralised particles. In conclusion, more work is necessary to confirm the effect of the tubular environment on extracellular matrix production by rat osteoblasts.

Clearly this type of culture has great potential for further investigation and possible applications in cell engineering. It may give us an insight into the behaviour of cells not seen in monolayer culture due to the artificial conditions of such a situation. Chondrocytes and osteoblasts clearly respond to the tube environment and further investigation of their response to this environment could ultimately help tissue repair *in vivo*.

References

Aranow, M., Gedrstenfield, L., Owen, T., *et al.* (1990) Factors that promote progressive development of the osteoblast phenotype in cultured fetal rat calvaria cells. *J. Cell. Physiol.* **143:** 213–221

Aubin, J.E., Liu, F., Malaval, L., Gupta, A.K. (1995) Osteoblast and chondroblast differentiation. *Bone* **17**[Suppl 2]: 77S–83S

Bobyn, J.D., Pilliar, R.M., Cameron, H.U., Weatherly, G.C. (1980) The optimum pore size for the fixation of porous-surfaced metal implants by the ingrowth of bone. *Clin. Orthop.* **150:** 263–270

Boyan, B.D., Hummert, T.W., Dean, D.D., Schwartz, Z. (1996) Role of material surfaces in regulating bone and cartilage cell response. *Biomaterials* **17:** 137–146

Brunette, D.M. (1988) The effects of implant surface topography on the behaviour of cells. *Int. J. Oral Maxillofac. Implants* **3:** 231–246

Brunette, D.M., Ratkay, J., Chehroudi, B. (1991) Behaviour of osteoblasts on micromachined surfaces. In: Davies, J.E. (ed.) *The Bone-Biomaterial Interface.* University of Toronto Press, Toronto

Chehroudi, B., Brunette, D.M. (1995) Effects of surface topography on cell behaviour. In: Wise, D.J.T.D.L., Altobelli, D.E., Yaszenski, M.J., Gresser, G.D., Schwartz, E.R. (eds.) *Encyclopedic handbook of biomaterials and bioengineering. Part S. Materials.* Dekker, New York, pp. 813–842

Chehroudi, B., Ratkay, J., Brunette, D.M. (1992) The role of implant surface geometry on mineralization in vivo and in vitro ó a transmission and scanning electron-microscopic study. *Cells Mater.* **2:** 89–104

Chehroudi, B., McDonnell, D., Brunette, D.M. (1997) The effects of micromachined surfaces on formation of bonelike tissue on subcutaneous implants as assessed by radiography and computer image processing. *J. Biomed. Mater. Res.* **34:** 279–290

Curtis, A.S.G., Varde, M. (1964) Control of cell behaviour: topological factors. *J. Natl. Cancer Inst.* **33:** 15–26

Curtis, A.S.G., Wilkinson, C.D.W. (1998) Topographical control of cells. *Biomaterials* (in press)

Dunn, G.A. (1982) Contact guidance of cultured tissue cells: a survey of potentially relevant properties of the substratum. In: Bellairs, R., Curtis A., Dunn, G. (eds.) *Cell behaviour.* Cambridge University Press, Cambridge, pp. 247–281

Dunn, G.A., Heath, J.P. (1976) A new hypothesis of contact guidance in tissue cells. *Exp. Cell Res.* **101:** 1–14

Eggli, P.S., Mueller, W., Schenk, R.K. (1987) The role of pore size on bone ingrowth and implant substitution in hydroxyapatite and tricalcium phosphate ceramics; a histologic and morphometric study in rabbits. In: Pizzoferrato, A., Marchetti, P.G., Ravaglioli, A., Lee, A.J.C.(eds.) *Biomaterials and clinical applications.* Elsevier Science, Amsterdam, pp. 53–56

Escarot-Charrier, B., Shepard, N., Charette, G., *et al.* (1988) Mineralization in osteoblast cultures: a light and electron microscopic study. *Bone* **9:** 147–154
Gray, C., Boyde, A., Jones, S.J. (1996) Topographically induced bone formation in vitro: implications for bone implants and bone grafts. *Bone* **18:** 115–123
Gundberg, C.M., Hauschka, P.V., Lian, J.B., Gallop, P.M. (1984) Osteocalcin: isolation, characterisation, and detection. *Methods Enzymol.* **107:** 516–545
Hung, C.T., Pollack, S.R., Reilly, T.M., Brighton, C.T. (1995) Real-time calcium response of cultured bone cells to fluid flow. *Clin. Orthop. Related Res.* **313:** 256–269
Qu, J., Chehroudi, B. Brunette, D.M. (1996) The use of micromachined surfaces to investigate the cell behavioural factors essential to osseointegration. *Oral Dis.* **2:** 102–115
Robey, P.G. (1989) The biochemistry of bone. *Endocrinol. Metab. Clin. North Am.* **18:** 859–901
Selye, H., Lemire, Y., Bajusz, E. (1960) Induction of bone, cartilage and hemopoietic tissue by subcutaneously implanted tissue diaphragms. *Roux Arch. Entwicklungsmechanik* **151:** 572–585
Twitty, V.C., Niu, M.C. (1954) The motivation of cell migration, studied by isolation of embryonic pigment cells singly and in small groups in vitro. *J. Exp. Zool,* **125:** 541–573
Von der Mark, K. (1986) Differentiation, modulation and dedifferentiation of chondrocytes. *Rheumatology* **10:** 272–315
Wojciak, B., Crossan, J., Curtis, A.S.G., Wilkinson, C.D.W. (1995) Grooved substrata facilitate in vitro healing of completely divided flexor tendons. *J. Materials Sci. Materials Med.* **6:** 266–271

Functional characterization of a muscarinic receptor in the smooth muscle of the shark (*Squalus acanthias*) ventral aorta

David. H. Evans[1] · Mark P. Gunderson[2]

[1]Department of Zoology, University of Florida, Gainesville, FL 32611-2009, USA
[2]Mt. Desert Island Biological Laboratory, Salsbury Cove, ME 04672, USA

Received: 19 January 1998 / Accepted: 13 March 1998 / Published: 23 March 1998

Abstract. A suite of muscarinic receptor blockers was used to characterize the receptor(s) mediating the contractile effect of acetylcholine (ACh) on isolated rings of ventral aorta from the dogfish shark, *Squalus acanthias*. The M_2/M_4-specific inhibitor *N,N'-bis*(6-{[(2-methoxyphenyl) methyl] amino}hexyl)-1,8-octane diamine tetrahydrochloride (methoctramine) did not reduce the efficacy of ACh, and the M_3-specific inhibitor 4-diphenylacetoxy-*N*-methyl-piperidine methiodide (4-DAMP) displaced the ACh concentration-response curve to the right at much lower concentrations than the M_1-specific inhibitor (5 - 11 - dihydro - 11 - [4 - methyl - l - piperazinyl) acetyl] - 6 *H* - pyrido [2,3-b] [1,4] benzodiazepin - 6 - one dihydrochloride) (pirenzepine). It appears, therefore, that an M_3-type muscarinic receptor is expressed in the aortic vascular smooth muscle of the dogfish shark.

Key words. Dogfish · Shark · Muscarinic receptor · *Squalus acanthias* Vascular smooth muscle · Ventral aorta

Introduction

The smooth muscle of fish blood vessels responds to acetylcholine (ACh) by contraction, producing an increase in tension in isolated vascular tissue (e.g., Small *et al.*, 1990; Olson and Villa, 1991; Miller and Vanhoutte, 1992; Farrell and Johansen, 1995; Evans and Gunderson, 1998) or hypertension in intact animals (e.g., Chan and Chow, 1976; Farrell, 1981; Lipke *et al.*, 1990). The contractile response is found even in the presence of an intact endothelium (e.g., Olson and Villa, 1991; Miller and Vanhoutte, 1992; Evans and Gunderson, 1998), which suggests that, contrary to those in mammals (e.g., Furchgott and Zawadzki,

Correspondence to: D.H. Evans (e-mail: dhefish@zoo.ufl.edu, Fax: +1-352-923704)

1980; Griffith *et al.*, 1984), fish vascular endothelial cells do not express ACh receptors coupled to an endothelium-derived relaxing factor (EDRF) system.

Despite contractile responses to ACh, it is not clear if there is cholinergic innervation of teleost systemic vasculature (Nilsson, 1983), but the branchial vasculature of at least the cod (*Gadus morhua*) has cholinergic nerves connected to the efferent filamental artery (Nilsson and Pettersson, 1981). There does not appear to be any cholinergic innervation of either the systemic or the branchial vasculature in the elasmobranchs (Nilsson, 1983; Metcalfe and Butler, 1984; Morris and Nilsson, 1994), although at least two studies have demonstrated the sensitivity of shark vessels (coronary artery and ventral aorta) to ACh (Farrell and Johansen, 1995; Evans and Gunderson, 1998).

Five unique muscarinic receptors (termed M_1–M_5) have been identified in functional or pharmacological studies of a variety of mammalian smooth muscle types, and a similar number of receptors have now been cloned (termed m1–m5; e.g., Hulme *et al.*, 1990; Caulfield, 1993; Eglen *et al.*, 1996; Ehlert *et al.*, 1997). ACh-induced dilation of mammalian vascular smooth muscle [mediated by endothelial receptors that stimulate the production of nitric oxide (NO)] is generally via the M_3 receptor; extant data support the possible involvement of M_1–M_4 receptors in contractile responses, but the M_3 receptor is most often implicated (e.g., Caulfield, 1993; Eglen *et al.*, 1996; Norel *et al.*, 1996).

Since we have found that the isolated vascular ring from the shark *Squalus acanthias* ventral aorta contracts when ACh is applied whether the endothelium is present or not (Evans and Gunderson, 1998), the current study was undertaken to characterize the type(s) of muscarinic receptor(s) expressed by the aortic vascular smooth muscle of this species.

Materials and methods

The collection and maintenance of *S. acanthias*, preparation and mounting of aortic rings, and recording of vascular tension have been described previously (Evans and Weingarten, 1990; Evans *et al.*, 1996). The lumen of the ring was rubbed gently with a roughened polyethylene tube (PE 90) to remove the intimal endothelial cells before mounting, and the rings were stabilized at 500 mg tension (4.91 mN) before each experiment. Tension was recorded via Gould-Statham strain transducers, whose output was recorded either by a Gilson Duograph or a Biopac MP100WS System, using AcqKnowledge III software, connected to a Macintosh Powerbook Model 140. The control ACh concentration-response curve was generated by the cumulative addition of ACh in distilled water to the experimental bath of elasmobranch Ringerís. The efficacy of putative muscarinic receptor antagonists was determined by generating ACh concentration-response curves in the presence of increasing concentrations of antagonist. In all protocols, including the initial ACh experiment, the rings were first exposed to 10^{-4} M ACh to ensure that the tissue was responsive, and

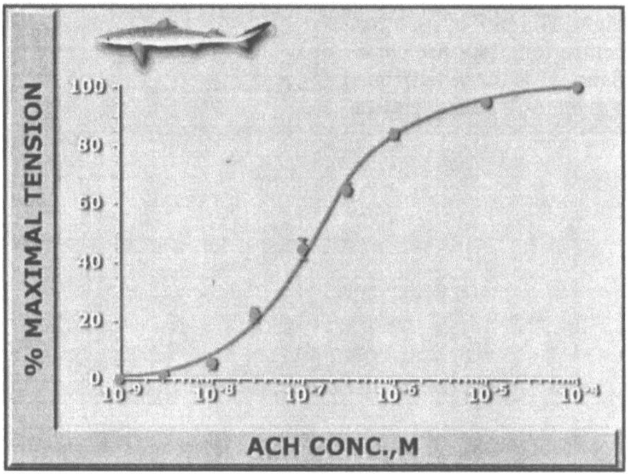

Fig. 1. Acetylcholine concentration-response curve of isolated aortic rings from the ventral aorta of the shark *Squalus acanthias*. The rings were free of endothelium. Data are mean ± SEM, n=4

because preliminary experiments demonstrated that this initial treatment maximized subsequent responses to ACh. The rings were then washed three times with elasmobranch Ringerís and allowed to re-stabilize before subsequent addition of either ACh or antagonist.

The muscarinic receptor antagonists pirenzepine (5-11-dihydro-11-[4-methyl-l-piperazinyl)acetyl]-6*H*-pyrido[2,3-b][1,4]benzodiazepin-6-one dihydrochloride), methoctramine [*N,N'*-*bis*(6-{[(2-methoxyphenyl) methyl] amino}hexyl)-1,8-octane diamine tetrahydrochloride], and 4-DAMP (4-diphenylacetoxy-*N*-methylpiperidine methiodide) (Research Biochemicals International) were dissolved in distilled water to a concentration of 10^{-2} to 10^{-4} M and then diluted in the 10-ml experimental bath to the desired concentration. None of the putative receptor antagonists reduced the final response to 10^{-4} M ACh ($p>0.05$), indicating that the effects were competitive (e.g., Bowman and Rand, 1980).

All data are presented as mean ± SE (n). The EC_{50} for the ACh-mediated contraction was computed from nonlinear regression of the concentration-response curve. Regression and statistical analyses were carried out using either Cricket Graph (1.3; Cricket Software) or Prism (2.0; GraphPad Software).

Results and discussion

Acetylcholine produced a concentration-dependent contraction of the isolated aortic ring from the ventral aorta of *S. acanthias* (Fig. 1) with an EC_{50} of 1.25×10^{-7} M, slightly (and significantly, p=0.014) greater than that found for this species in another series of experiments (7.85×10^{-8} M; Evans and Gunderson, 1998).

Transformation of the data (between 3×10^{-8} and 10^{-6} M) to a Hill Plot (Fig. 2; Bowman and Rand, 1980) generated a slope (apparent ligand/receptor) of 0.829 (r^2=1.00), which suggests that a single receptor mediates the response to ACh, although we cannot rule out two receptors.

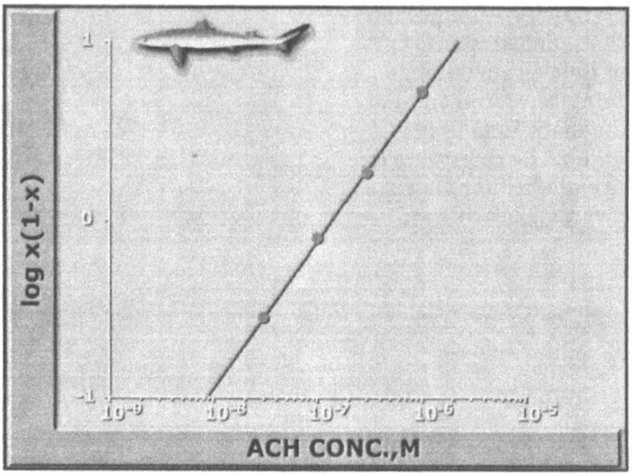

Fig. 2. Hill plot of ACh concentration-response curve data. „x“ is the % Sat/100 at a given ACh concentration

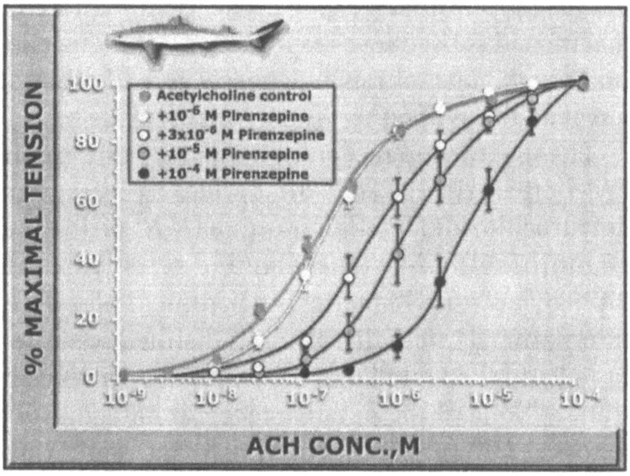

Fig. 3. The effect of pirenzepine on the ACh concentration-response curve. n=4–5

The M_1-selective antagonist pirenzepine (Hammer *et al.*, 1980; Eglen *et al.*, 1996) produced a rightward-shift in the ACh concentration-response curve (Fig. 3) but only at moderately high concentrations.

The M_2- and M_4-selective antagonist methoctramine (Melchiorre *et al.*, 1987; Eglen *et al.*, 1996) did not reduce the efficacy of ACh, even at 10^{-5} M (Fig. 4).

However, the M_3-selective antagonists, 4-DAMP (Eglen *et al.*, 1996) produced a significant rightward-shift of the ACh concentration-response curve with a relatively high efficacy (Fig. 5a).

If one plots the log of the ratio (x) of the concentration of ACh which produced equivalent responses in the presence and absence of the putative antagonist (minus 1) against the antagonist concentration, a Schild plot is generated (Bowman and Rand, 1980) which can be used to compare the efficacy of antagonists. The x-intercept of such plots is the pA_2, the –log of the antagonist concentration that produces a 50% reduction in the efficacy of ACh.

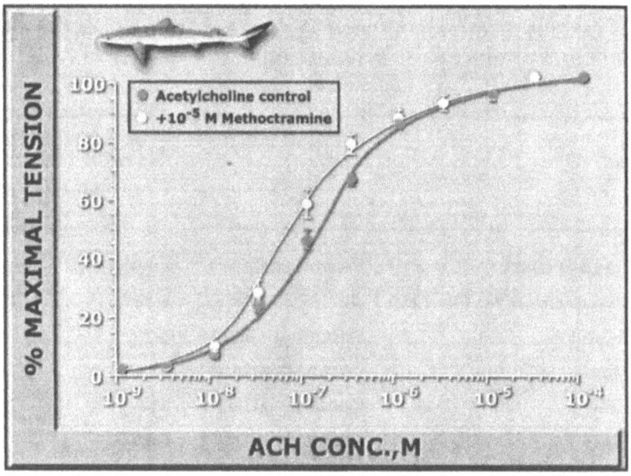

Fig. 4. The effect of 10^{-5} M methoctramine on the ACh concentration-response curve. n=6

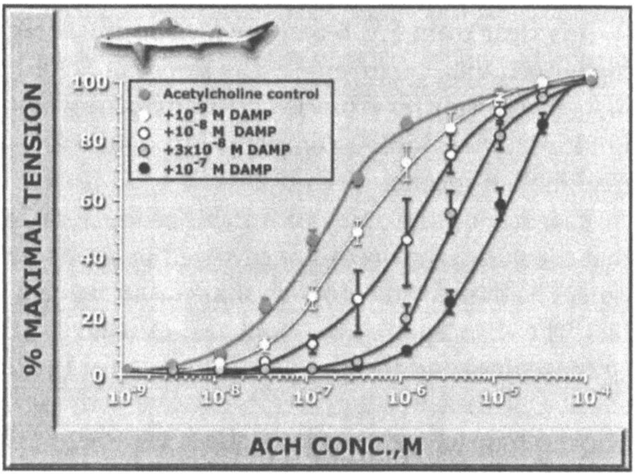

Fig. 5. The effect of 4-DAMP (**a** n=4) on the ACh concentration-response curve

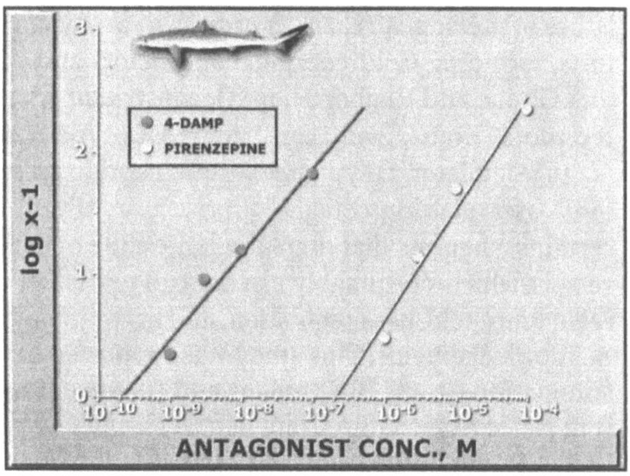

Fig. 6. Schild Plot of the effect of pirenzepine and 4-DAMP the ACh concentration-response curve

Table 1. A comparison of Schild plot data from this study and published data from studies of various mammalian species Values are the pA_2, i.e., the –log of the antagonist concentration that produces a 50% reduction in the efficacy of ACh

Species	Vessel	Antagonist pA_2 value		Reference
		Pirenzepine	4-DAMP	
Dogfish shark	Ventral aorta	6.69	9.67	this paper
Hypertensive rat (spontan.)	Aorta	6.74	9.39	Boulanger *et al.*, 1994
Rabbit	Pulmonary artery	6.77	8.92	Altiere *et al.*, 1994
Lamb	Coronary artery	7.49		Simonsen *et al.*, 1993
Cow	Coronary artery	7.14	9.04	Brunner *et al.*, 1991
Pig	Coronary artery	6.98	9.20	Entzeroth *et al.*, 1990

It is clear from Fig. 6 that the M_3-specific antagonist 4-DAMP has a much higher pA_2 than pirenzepine. The slopes of Schild plots for these two antagonists did not differ from 1.0, demonstrating competitive interaction (e.g., Limbird, 1986). Table 1 compares our data with some published for various mammalian species.

Based upon the array of pA_2, all of these mammalian studies concluded that the muscarinic receptor involved in the contractile response was the M_3 type. The data for the dogfish shark ventral aorta fall into the same array; in fact, the discrepancy between the efficacy of the M_1-specific antagonist (pirenzepine) and the M_3-specific antagonist (4-DAMP) is the greatest for the shark. Our data are, therefore, consistent with the hypothesis that the muscarinic receptor on the vascular smooth muscle of the shark ventral aorta is also the M_3 type.

We have now demonstrated that the ventral aorta and/or branchial vasculature of the dogfish shark expresses a variety of receptors for vasoactive agonists, including α-adrenergic (constrictor) and β-adrenergic (dilator) receptors (Evans and Claiborne, 1983), constrictor (A_1) and dilator (A_2) adenosine receptors (Evans, 1992), constrictor, ET_B-type endothelin receptors (Evans *et al.*, 1996), dilator, C-type natriuretic peptide receptors (NPR-B; (Evans *et al.*, 1993), and dilator prostaglandin receptors (Evans and Gunderson, 1998). This certainly suggests that there is a large suite of control sites expressed in these vessels, which presumably can control perfusion of the gill, the site of osmoregulation, acid-base regulation, and excretion of nitrogen (e.g., Evans, 1998).

Acknowledgements. This study was supported in part by National Science Foundation Grants IBN9306997 and IBN-9604824; the Maine Affiliate, American Heart Association Grant 9507715S; and NIEHS Grant P30-ESO3238 to the Center for Membrane Toxicity Studies at the MDIBL. Appreciation is expressed to Ms. Laurie Walz, who did much more than merely render our PICT files into JPEG and GIF. The artistry is hers.

References

Altiere, R.J., Travis, D.C., Roberts, J., Thompson, D.C. (1994) Pharmacological characterization of muscarinic receptors mediating acetylcholine-induced contraction and relaxation in rabbit intrapulmonary arteries. *J. Pharmacol. Exp. Ther.* **270**: 269–276

Boulanger, C.M., Morrison, K.J., Vanhoutte, P.M. (1994) Mediation by M_3-muscarinic receptors of both endothelium-dependent contraction and relaxation to acetylcholine in the aorta of the spontaneously hypertensive rat. *Br. J. Pharmacol.* **112**: 519–524

Bowman, W.C., Rand, M.J. (1980) *Textbook of pharmacology*. Blackwell Scientific, London

Brunner, F., Kuhberger, E., Schloos, J., Kukovetz, W.R. (1991) Characterization of muscarinic receptors of bovine coronary artery by functional and radioligand binding studies. *Eur. J. Pharmacol.* **196**: 247–255

Caulfield, M.P. (1993) Muscarinic receptors - characterization, coupling and function. *Pharmacol. Ther.* **58**: 319–379

Chan, D.K.O., Chow, P.H. (1976) The effects of acetylcholine, biogenic amines and other vasoactive agents on the cardiovascular functions of the eel, *Anguilla japonica. J. Exp. Zool.* **196**: 13–26

Eglen, R.M., Hegde, S.S., Watson, N. (1996) Muscarinic receptor subtypes and smooth muscle function. *Pharmacol. Rev.* **48**: 531–565

Ehlert, F.J., Ostrom, R.S., Sawyer, G.W. (1997) Subtypes of the muscarinic receptor in smooth muscle. *Life Sci.* **61**: 1729–1740

Entzeroth, M., Doods, H.N., Mayer, N. (1990) Characterization of porcine coronary muscarinic receptors. *Naunyn-Schmiedebergs Arch. Pharmacol.* **341**: 432–438

Evans, D.H. (1992) Evidence for the presence of A_1 and A_2 adenosine receptors in the ventral aorta of the dogfish shark, *Squalus acanthias. J Comp Physiol [B]* **162**: 179–183

Evans, D.H. (1998) *The physiology of fishes*, 2nd Edn. CRC, Boca Raton, Fla., pp. 1–519

Evans, D.H., Claiborne, J.B. (1983) Haemodynamic effects of adrenaline on the isolated, perfused head of the dogfish „pup" (*Squalus acanthias*) *J. Exp. Biol.* **105**: 363–371

Evans, D.H., Gunderson, M.P. (1998) A prostaglandin, not nitric oxide, mediates endothelium-dependent dilation in the ventral aorta of the shark *Squalus acanthias. Am. J. Physiol.* **274** (in press)

Evans, D.H., Weingarten, K.E. (1990) The effect of cadmium and other metals on vascular smooth muscle of the dogfish shark, *Squalus acanthias. Toxicology* **61**: 275–281

Evans, D.H., Toop, T., Donald, J., Forrest, J.N.J. (1993) C-type natriuretic peptides are potent dilators of shark vascular smooth muscle. *J. Exp. Zool.* **265**: 84–87

Evans, D.H., Gunderson, M., Cegelis, C. (1996) ET_B-type receptors mediate endothelin-stimulated contraction in the aortic vascular smooth muscle of the spiny dogfish shark, *Squalus acanthias*. *J. Comp. Physiol.* **165**: 659–664

Farrell, A.P. (1981) Cardiovascular changes in the lingcod (*Ophiodon elongatus*) following adrenergic and cholinergic drug infusions. *J. Exp. Biol.* **91**: 293–305

Farrell, A.P., Johansen, J.A. (1995) Vasoactivity of the coronary artery of rainbow trout, steelhead trout, and dogfish: lack of support for non prostanoid endothelium-derived relaxation factors. *Can. J. Zool.* **73**: 1899–1911

Furchgott, R.F., Zawadzki, J.V. (1980) The obligatory role of the endothelial cells in the relaxation of arterial smooth muscle by acetylcholine. *Nature* **288**: 373–376

Griffith, T.M., Henderson, A.H., Edwards, H.D., Lewis, M.J. (1984) Isolated perfused rabbit coronary artery and aortic strip preparations: the role of endothelium-derived relaxant factor. *J. Physiol. (Lond.)* **351**: 13–24

Hammer, R., Berrie, C.P., Birdsall, N.J.M., Burgen, A.S.V., Hulme, E.C. (1980) Pirenzepine distinguishes between different subclasses of muscarinic receptors. *Nature (Lond.)* **238**: 90–92

Hulme, E.C., Birdsall, N.J.M., Buckley, N.J. (1990) Muscarinic receptor subtypes. *Annu. Rev. Pharmacol. Toxicol.* **30**: 633–673

Limbird, L.E. (1986) *Cell surface receptors: a short course on theory and methods*. Martinus Nijhoff, Boston, pp. 1–196

Lipke, D.W., Oparil, S., Olson, K.R. (1990) Vascular effects of kinins in trout and bradykinin metabolism by perfused gill. *Am. J. Physiol.* **258**: R515–R522

Melchiorre, C., Angeli, P., Lambrecht, G., Mutschler, E., Micchio, M.R., Wess, J. (1987) Antimuscarinic action of methoctramine, a new cardioselective M-2 selective muscarinic receptor antagonist, alone and in combination with atropine and gallamine. *Eur. J. Pharmacol.* **144**: 117–124

Metcalfe, J.D., Butler, P.J. (1984) On the nervous regulation of gill blood flow in the dogfish (*Scyliorhinus canicula*) *J. Exp. Biol.* **113**: 253–268

Miller, V.M., Vanhoutte, P.M. (1992) Endothelium-dependent vascular responsiveness: evolutionary aspects. In: Ryan, U.S., Rubanyi, G.M. (eds.) *Endothelial regulation of vascular tone*. Dekker, New York, pp. 3–20

Morris, J.L., Nilsson, S. (1994) The circulatory system. In: Nilsson, S., Holmgren, S. (eds.) *Comparative physiology and evolution of the autonomic nervous system*. Harwood Academic, Chur, pp. 193–246

Nilsson, S. (1983) *Autonomic nerve function in the vertebrates*. Springer, Berlin Heidelberg New York, pp. 1–253

Nilsson, S., Pettersson, K. (1981) Sympathetic nervous control of blood flow in he gills of the Atlantic cod, *Gadus morhua*. *J. Comp. Physiol.* **144**: 157–163

Norel, X., Walch, L., Costantino, M., Labat, C., Gorenne, I., Dulmet, E., Rossi, F., Brink, C. (1996) M_1 and M_3 muscarinic receptors in human pulmonary arteries. *Br. J. Pharmacol.* **119**: 149–157

Olson, K.R., Villa, J. (1991) Evidence against nonprostanoid endothelium-derived relaxing factor(s) in trout vessels. *Am. J. Physiol.* **260**: R925–R933

Simonsen, U., Prieto, D., Rivera, L., Hernandez, M., Mulvany, M.J., Garcia-Sacristan, A. (1993) Heterogeneity of muscarinic receptors in lamb isolated coronary resistance arteries. *Br. J. Pharmacol.* **109**: 998–1007

Small, S.A., MacDonald, C., Farrell, A.P. (1990) Vascular reactivity of the coronary artery in rainbow trout (*Oncorhynchus mykiss*). *Am. J. Physiol.* **258**: R1402–R1410

Deciphering the role of 14-3-3 proteins

Michael Gjedde Palmgren · Anja Thoe Fuglsang · Thomas Jahn

Plant Physiology and Anatomy Laboratory, The Royal Veterinary and Agricultural University, Thorvaldsensvej 40, 1871 Frederiksberg C, Copenhagen, Denmark

Received: 3 March 1998 / Accepted: 16 April 1998 / Published: 26 May 1998

Abstract. 14-3-3 Proteins are found to bind to a growing number of eukaryotic proteins and evidence is accumulating that 14-3-3 proteins serve as modulators of enzyme activity. Several 14-3-3 protein recognition motifs have been identified and an increasing number of target proteins have been found to contain more than one binding site for a 14-3-3 protein. It is thus possible that 14-3-3 dimers function as clamps that simultaneously bind to two motifs within a single binding partner. Phosphorylation of a number of binding motifs has been shown to increase the affinity for 14-3-3 proteins but other mechanisms also regulate the association. It has recently been demonstrated that fusicoccin induces a tight association between 14-3-3 proteins and the plant plasma membrane H^+-ATPase. Phorbol esters and other hydrophobic molecules may have a similar effect on the association between 14-3-3 proteins and specific binding partners.

Key words. Binding motif · H^+-ATPase · Plasma membrane · 14-3-3 Protein
Protein kinase C · Modulator · Raf-1 · Signal transduction

Introduction

The 14-3-3 proteins were identified in the late 1960s as being abundantly present in brain tissue (Moore and Perez 1967). Their function was, however, unknown, and it was not until recently that they were „rediscovered" when they were found to play important roles in cell cycle regulation, differentiation, exocytosis, targeting of proteins to various locations in the cell and in the coordination of signal transduction pathways (for reviews, see Aitken 1996; Ferl 1996). Mutants in 14-3-3 genes are beginning to illuminate some of the specific functions of 14-3-3 proteins in various cell types (Chang and Rubin,

Correspondence to: M. G. Palmgren (e-mail: palmgren@biobase.dk, Fax: +45-3528-3333)

1997; Kockel *et al.* 1997; Roberts *et al.* 1997). How 14-3-3 proteins carry out a diverse array of functions is still enigmatic. Although most organisms have a number of 14-3-3 isoforms, their primary structures are extremely well conserved, suggesting a common molecular mechanism of action.

14-3-3 proteins bind to several target proteins

One feature is common to all 14-3-3 proteins: they bind to other proteins. Table 1 lists a number of proteins that interact with 14-3-3 proteins. Different methods have been employed to show interaction, such as the yeast two-hybrid system, immunoprecipitation and co-purification. Interactions have been demonstrated both *in vivo* and in vitro. The notion that 14-3-3 proteins form complexes with other proteins therefore seems to be well based.

Isoforms of 14-3-3 proteins show little specificity

All tissues in all eukaryotic organisms analysed so far possess 14-3-3 proteins (Wang and Shakes 1996). Only one 14-3-3 protein has been found in the amoebae *Dictyostelium discoideum* (Knetsch *et al.* 1997), two in the yeast *Saccharomyces cerevisiae* (van Heusden *et al.* 1995), and at least ten and seven in *Arabidopsis* (Wu *et al.* 1997a) and humans (Aitken 1996), respectively. It has been suggested that each 14-3-3 isoform has a special function in the cell, with a selectivity towards a specific binding partner. If this concept reflected the true situation, it would explain how a spectrum of 14-3-3 proteins could play multiple roles.

There is, however, little indication that such isoform specificity exists. A single 14-3-3 isoform can bind to several binding partners *in vitro* and exert multiple biological functions (Lu *et al.* 1994). In addition, all those isoforms of *Arabidopsis* 14-3-3 protein that could be expressed in yeast complement the corresponding yeast genes (van Heusden *et al.* 1996).

In plants, 14-3-3 proteins have been implicated in the binding of fusicoccin (de Boer 1997; Fullone *et al.* 1998), a wilt-inducing toxin produced by the fungus *Fusicoccum amygdalii.* Fusicoccin binding sites are absent from fungi and animals (Meyer *et al.* 1993), suggesting that a specific function of one or more plant 14-3-3 proteins is to bind fusicoccin. However, it has recently been demonstrated that the fusicoccin binding site is a complex between two proteins: a 14-3-3 protein and the C-terminal regulatory domain of plant membrane H^+-ATPase (Baunsgaard *et al.* 1998). All tested plant, animal and yeast 14-3-3 proteins could form fusicoccin binding sites provided that plant plasma membrane H^+-ATPase was present, indicating that specificity is generated by the H^+-ATPase, and not by any of the 14-3-3 proteins (Baunsgaard *et al.* 1998). Post-translational modifications affect the potency of 14-3-3 proteins (Aitken *et al.* 1995), which could explain some of the quantitative differences often observed between isoforms.

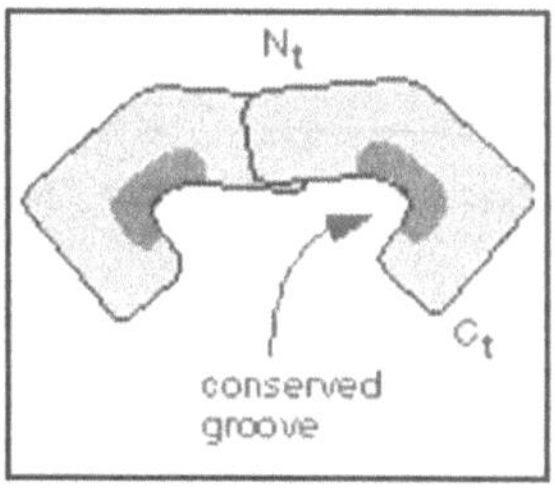

Fig. 1. Schematic presentation of the structure of 14-3-3 protein. Two monomers interact with each other at their N-termini to form a dimer in the crystal structure (Liu *et al.* 1995; Xiao *et al.* 1995). The groove of each monomer is almost invariant between 14-3-3 proteins and forms a binding site for a sequence motif present in a number of targets known to bind 14-3-3 proteins (Yaffe *et al.* 1997). The outer surface of the 14-3-3 protein is less conserved and has also been suggested to play a role in 14-3-3 protein interactions with other proteins (Jones *et al.* 1995)

The crystal structures of two 14-3-3 proteins have been solved (Liu *et al.* 1995; Xiao *et al.* 1995). Monomers interact with each other at their N-termini to form dimers. A negatively charged groove located in the centre of each 14-3-3 monomer has an interior made up of extremely well-conserved amino acid residues. In contrast, the outer part of the 14-3-3 protein is covered with amino acid side chains that are rather variable (Fig. 1). It has been suggested that this less conserved part of the molecule is responsible for the interaction with target proteins, since structural variation could determine specificity towards various binding partners (Aitken 1996; Jones *et al.* 1995). However, a 14-3-3 protein has been co-crystallized with a peptide representing a common 14-3-3 protein binding motif (RXXpSXP; see below) and the peptide found to interact with the groove (Yaffe *et al.* 1997). It is not known whether other binding motifs (see below) bind to the same site of the 14-3-3 protein. However, the conserved nature of the peptide binding site supports the notion that 14-3-3 proteins have very little specificity in binding to target proteins.

14-3-3 Isoforms often exhibit a distinct developmental and tissue-specific pattern of localization (Daugherty *et al.* 1996; Wang and Shakes 1997; Watanabe *et al.* 1991, 1993a, b, 1994). This controlled expression suggests a genetic, rather than a functional, model for 14-3-3 isoform diversity: the important differences between 14-3-3 genes are found in the promoter regions. The existence of multiple isoforms is thus a means to provide each cell in a multicellular organism with its specific requirement for 14-3-3 proteins at any time during development.

Do 14-3-3 proteins act as adapter proteins?

14-3-3 Proteins form dimers under native conditions. Homo- as well as heterodimer formation has been shown to occur both *in vivo* and *in vitro* (Jones *et al.* 1995; Wu *et al.* 1997b). These results imply that each 14-3-3 protein dimer can simultaneously bind to two different target proteins. Accordingly, a model has been postulated in which the 14-3-3 dimer functions as a scaffold that brings two target proteins into close proximity (Fig. 2A) (Aitken 1996; Braselmann and McCormick 1995; Xiao *et al.* 1995). There are, however, a number of arguments against this possibility: (1) except in the case of struc-

Table 1. Proteins interacting with 14-3-3 proteins

Partner of interaction	Methods[a]	Molecular activity of 14-3-3	References
A20	b, c	Chaperone	de Valck *et al.* 1997; Vincenz and Dixit 1996
Ascorbate, peroxidases	c		Zhang H *et al.* 1997
BAD	c, d	Inhibitor	Hsu *et al.* 1997; Zha *et al.* 1996
Breakpoint cluster region protein (BCR)	b, c, e, f	Adapter	Braselmann and McCormick 1995; Ichimura *et al.* 1997; Reuther *et al.* 1994
cdc25 phosphatase	c, h	Inhibitor	Conklin *et al.* 1995; Kumagai *et al.* 1998; Li and Sedivy 1993
cdk PCTAIRE-1	c, d	–	Sladeczek *et al.* 1997
Proto-oncogene product c-Clb	b, c, d	–	Liu *et al.* 1996; Robertson *et al.* 1997
Exoenzyme S	d	Activator	Fu *et al.* 1993; Zhang L *et al.* 1997
G-box binding factor	b	–	de Vetten *et al.* 1992
Glucocorticoid receptor (GR)	c, d	Activator	Wakui *et al.* 1997
Glycoprotein Ib-IX	b, e	–	Andrews *et al.* 1998; Du *et al.* 1996
Type 1 insulin growth factor receptor (IGFIR)	c	–	Craparo *et al.* 1997; Furlanetto *et al.* 1997
Insulin receptor substrate 1	b, c, d	–	Craparo *et al.* 1997; Kosaki *et al.* 1998; Ogihara *et al.* 1997
43-kDa Inositol polyphosphate 5-phosphatase	d, e	–	Campbell *et al.* 1997
Keratin intermediate filament	d	–	Liao and Omary 1996
Kinase suppressor of Ras (KSR)	b, c, d	–	Xing *et al.* 1997
NADH: nitrate reductase (NR)	e	Inhibitor	Bachmann *et al.* 1996; Moorhead *et al.* 1996
Nonstructural protein NS2	b	–	Brockhaus *et al.* 1996
Plasma membrane H^+-ATPase	b, c	Activator	Jahn *et al.* 1997; Oecking *et al.* 1997

Partner of interaction	Methods[a]	Molecular activity of 14-3-3	References
Polyoma virus middle T antigen	e	–	Cullere *et al.* 1998; Yaffe *et al.* 1997
Presequence region of mitochondrial precursor proteins	d	Chaperone	Alam *et al.* 1994
Protein kinase C (PKC)	d, e	Uncertain	Acs *et al.* 1995; Aitken *et al.* 1995; Robinson *et al.* 1994
Protein kinase C theta (PKC theta)	b, d	Inhibitor	Meller *et al.* 1996
Myosin II heavy chain specific protein kinase C (MHC-PKC)	b	Inhibitor	Matto-Velin *et al.* 1997
Protein-tyrosine phosphatase 1(PTPH1)	c, d	–	Zhang SH *et al.* 1997
B-Raf	c, e	–	Muslin *et al.* 1996; Papin *et al.* 1996
c-Raf-1	d, e, h	Uncertain	Clark *et al.* 1997; Li *et al.* 1995; Morrison *et al.* 1993; Muslin *et al.* 1996
Tryptophan hydroxylase	b, d	Activator	Banik *et al.* 1997; Furukawa *et al.* 1993; Ichimura *et al.* 1995
vav Proto-oncogene product	g	–	Suen *et al.* 1995
Wee1 kinase	c, d	–	Honda *et al.* 1997

[a] Methods employed to demonstrate interaction
[b] Co-purification/co-immunoprecipitation
[c] Two-hybrid screen
[d] *In vitro* interaction (recombinant proteins/purified proteins)
[e] Peptide binding study
[f] Baculovirus expression
[g] Bacteriophage expression library
[h] *In vivo* study in *Xenopus laevis* oocytes

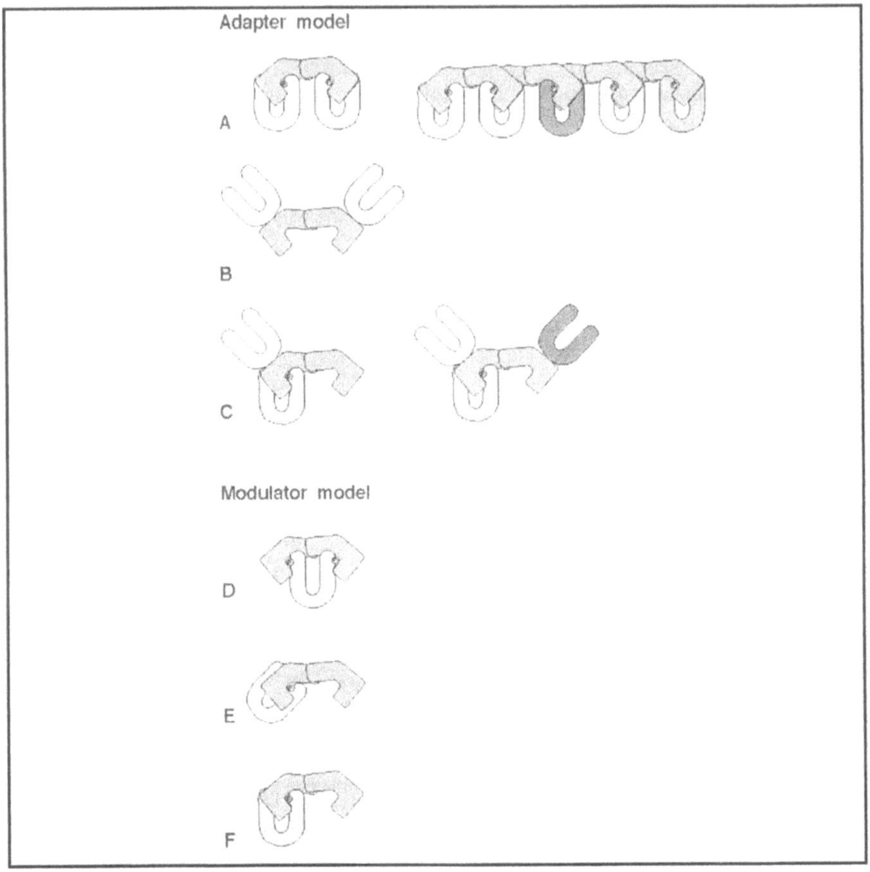

Fig. 2A–E. Models for interaction of 14-3-3 with target proteins. 14-3-3 Proteins have been suggested to function as adapters (Adaptor model; **A–C**), that link different proteins together, or as direct modulators of enzyme activity (Modulator model; **D,E**). According to the former model, binding of 14-3-3 protein is not sufficient to alter enzyme activity. Rather, the role of 14-3-3 protein is to bridge two 14-3-3 binding proteins, the enzyme and a regulatory protein. According to the latter model, binding of 14-3-3 protein stabilizes a particular conformation of the target protein. Release of the 14-3-3 protein would result in an alternative conformation of the protein having altered properties. *Small circles* indicate sites phosphorylated in the target protein. **A** Interaction of the target proteins at a phosphorylated binding motif with the highly conserved grooves of the 14-3-3 protein dimer (*left*). Given the situation that one or more target proteins have more than just one site for interaction, 14-3-3 proteins are likely to precipitate their target proteins (*right*). **B** Interaction of target proteins with the rather variable outer surface of the 14-3-3 protein dimer. **C** Interaction of target proteins with the 14-3-3 protein dimer at different sites: one protein interacts at the outer surface and the second protein binds to the groove structure of a 14-3-3 protein (*left*). The 14-3-3 protein dimer would allow for the link of more than just two different proteins (*right*). **D** Both sites at the target molecule interact with the groove structure of the 14-3-3 dimer. **E** The outer surface of the 14-3-3 protein is involved in the interaction. **F** Alternatively, modulation of activity is brought about by the interaction of a 14-3-3 protein at a single binding site of the target protein tural proteins, there is no documented need for an adapter to help bring together proteins in the cell; (2) given the large number of 14-3-3 protein interactions, the binding partner in the 14-3-3 dimer is likely to vary considerably, and thus presentation of a binding partner is not likely to be very specific; (3) the fact that a number of 14-3-3 binding proteins contain more than a single 14-3-3 binding site (see below) suggests that if 14-3-3 proteins serve adapter functions, they should at least in theory be able to form a complex protein network within the cell and possibly precipitate the 14-3-3 binding proteins in a fashion resembling that of polyclonal antibodies (Fig. 2A).

It remains still a theoretical possibility that interaction of target proteins with the more variable outer surface of the 14-3-3 dimer is involved in complexing proteins (Figs. 1and 2B) (Jones *et al.* 1995). In this interaction binding might in fact be more specific, and a 14-3-3 dimer, which then possesses a total of four binding sites, could actually bring more than just two proteins together (Fig. 2C). Experimental evidence that the outer surface of 14-3-3 proteins is involved in binding to proteins other than 14-3-3 is still lacking.

A general function of 14-3-3 proteins may be to serve as direct modulators of enzyme activity

Increasingly, 14-3-3 proteins are found to modulate the properties of other proteins (Sehnke and Ferl 1996; Table 1). The effect of 14-3-3 proteins can be either inhibitory or activatory. Alternatively, they can induce or reduce the binding of the target protein to other proteins.

14-3-3 Proteins activate proteins such as tyrosine and tryptophan hydroxylases (Ichimura *et al.* 1988), exoenzyme S (Fu *et al.* 1993), the glucocorticoid receptor (Wakui *et al.* 1997) and the plasma membrane H^+-ATPase (Baunsgaard *et al.* 1998; Jahn *et al.* 1997; Oecking *et al.* 1997). In contrast, it is inhibition that characterizes the interaction between 14-3-3 proteins and, for example, nitrate reductase (Bachmann *et al.* 1996; Moorhead *et al.* 1996).

The effect of 14-3-3 proteins on two important molecules in the regulation of cellular processes, protein kinase C and Raf-1, is more controversial. 14-3-3 Proteins have been reported to either (1) stimulate (Acs *et al.* 1995; Dellambra *et al.* 1995; Isobe *et al.* 1992; Tanji *et al.* 1994) or (2) inhibit (Meller *et al.* 1996; Robinson *et al.* 1994; Toker *et al.* 1992) protein kinase C. In addition, 14-3-3 proteins have been reported to either (1) activate (Fantl *et al.* 1994; Freed *et al.* 1994; Irie *et al.* 1994; Li *et al.* 1997), (2) inhibit (Clark *et al.* 1997; Rommel *et al.* 1997) or (3) have no influence on (Michaud *et al.* 1995; Suen *et al.* 1995) Raf-1.

Additional roles of 14-3-3 proteins are to solubilize proteins such as A20 (Vincenz and Dixit 1996) and keratins (Liao and Omary 1996) and to stimulate protein import into mitochondria (Alam *et al.* 1994).

Several 14-3-3 protein target sequences have been identified

A number of 14-3-3 binding motifs in target proteins have been described. A peptide derived from the Raf-1 sequence was shown to bind to a 14-3-3 protein *in vitro* provided that a serine in the sequence was phosphorylated (Muslin *et al.* 1996). The 14-3-3 binding motif in Raf-1 was found to be RSXpSXP (pS indicates a phosphorylated serine) and has been identified in several other 14-3-3 binding proteins (Muslin *et al.* 1996). Systematic screening of peptide libraries has identified another motif, RXY/FXpSXP (Yaffe *et al.* 1997) that binds to 14-3-3 proteins. In c-cbl an RX1-2SX2-3S 14-3-3 binding motif has been reported (Liu *et al.* 1997). A systematic analysis of the 14-3-3 protein-binding capacity of peptides derived from the glycoprotein Ib-IX-V complex revealed the presence of multiple sites for 14-3-3 protein binding, among which were GHSL and RLSLTDP (Andrews *et al.* 1998). On the basis of these results, it was suggested that RXSX(S/T)XP, a motif conserved in proteins with SH2 domains, is recognized by 14-3-3 proteins (Andrews *et al.* 1998). A 14-3-3 protein binding motif in the platelet adhesion receptor was suggested to be X3SX2SX3SX2SX (Du *et al.* 1996). However, this sequence in the platelet adhesion receptor includes the 14-3-3 protein binding motif GHSL (Andrews *et al.* 1998) which may represent the true 14-3-3 binding site in this protein.

The cysteine-rich domains (CRDs) of Raf-1, MHC-PKC and PKC have been implicated in the binding of 14-3-3 proteins. There is very little sequence similarity between these domains except for the motif CXTVK(P)C (Table 2) which is present in most of these CRDs. This motif might therefore be involved in interaction with the 14-3-3 protein.

Table 2. Sequence motifs involved in binding of 14-3-3 proteins

Sequence motif[a]	References
RSxSpSxP	Yaffe *et al.* 1997
RxYxpSxP	Yaffe *et al.* 1997
RxFxpSxP	Yaffe *et al.* 1997
RSxpS	Ogihara *et al.* 1997
SxpSxP	Ogihara *et al.* 1997
RSxpSxP	Muslin *et al.* 1996; Zha *et al.* 1996
RxRxxpS	Zha *et al.* 1996
RxSxpSxP	Andrews *et al.* 1998
Rx(x)Sxx(x)S	Liu *et al.* 1997
GHSL(P)	Andrews *et al.* 1998
CxTKV(P/x)xxC	this study

[a] pS denotes a phosphorylated serine

The ligand-binding domain of the glucocorticoid receptor (Wakui *et al.* 1997), which binds 14-3-3 proteins, does not have the aforementioned sequence motifs, implying that additional 14-3-3 protein-binding motifs exist.

It should be noted that the sequence of the synthetic peptide shown to have the highest affinity for 14-3-3 proteins in the study by Yaffe *et al.* (1997) does not make a perfect match to any protein sequence in the databases. Flexibility in binding motifs may ensure that binding is not too strong, thus allowing for more flexibility in regulation.

14-3-3 Proteins bind to multiple sites in several target proteins

Interestingly, a number of 14-3-3 binding proteins contain more than one 14-3-3 binding site (Table 3). These include Raf-1 (Muslin *et al.* 1996), c-cbl (Liu *et al.* 1997), glycoprotein Ib-IX-V complex (Andrews *et al.* 1998) and the glucocorticoid receptor (Wakui *et al.* 1997). This immediately suggests a mechanism by which 14-3-3 proteins could regulate the properties of their binding partners (Aitken 1996). According to this mechanism, each monomer in the 14-3-3 dimer binds to a distinct domain in the target protein (Fig. 2D, E). By binding to both domains simultaneously, the 14-3-3 dimer functions as a clamp that induces a conformational change in the target molecule. The conformational change may result in either activation or inhibition of target enzyme activity, or have other effects, such as the exposure or disclosure of important domains (Fig. 2D–F). From this model it follows that binding of the 14-3-3 protein to a single site in the target protein does not necessarily influence its conformational state, depending on the protein in question. This may explain why binding of 14-3-3 proteins to target proteins often does not result in a change in their activity (Fig. 3). In other proteins, binding of 14-3-3 protein to a single site might be sufficient to bring about a conformational change.

Reversible phosphorylation regulates the interaction between 14-3-3 and many target sequences

Interaction of 14-3-3 proteins with their target proteins is often facilitated by phosphorylation of the binding partner (Fig. 3). Tryptophan hydroxylase forms a complex with 14-3-3 proteins, in a process that requires phosphorylation of the former (Furukawa *et al.* 1993), whereas 14-3-3 proteins have little effect on the non-phosphorylated protein. Phosphorylation itself does not affect the activity of tryptophan hydroxylase, but allows for binding of the 14-3-3 protein, which in turn leads to activation (Furukawa *et al.* 1993). Purified target proteins may be dephosphorylated unless precautions have been taken; this may explain why 14-3-3 proteins often do not affect the activities of their target protein when they are mixed *in vitro* (Honda *et al.* 1997).

Phosphorylation of the 14-3-3 binding motifs often strongly affects the affinity for 14-3-3 protein. Phosphorylation of the serine in the RSXpSXP motif

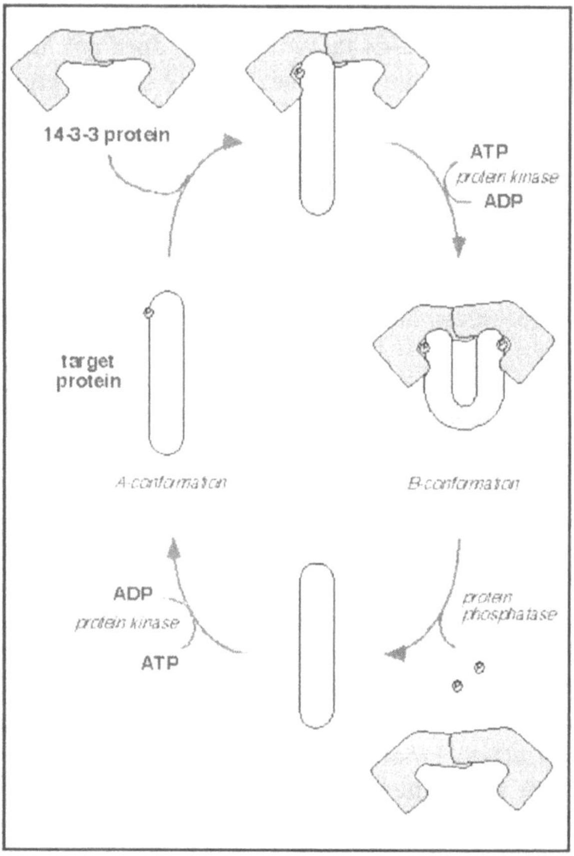

Fig. 3. Regulation of a target protein by 14-3-3 protein, protein kinase(s) and protein phosphatase(s). The target protein exists in two conformations, *A* and *B*, having different activity states. According to this model, the dimer of a 14-3-3 protein stabilizes a particular conformation (B-conformation) of the target protein. The shift from the original A-conformation to the B-conformation is only induced when the 14-3-3 protein binds simultaneously to two different sites of the target protein. *Small circles* indicate sites phosphorylated in the target. Interaction between the 14-3-3 protein and the target protein at a single site does not change the conformational state of the target protein. Furthermore, binding of 14-3-3 protein is increased by phosphorylation of 14-3-3 binding motifs in the target protein. Thus, interaction between 14-3-3 protein and its target can be closely regulated by pairs of protein kinases and phosphatases. Interactions between 14-3-3 proteins and their targets are not always dependent upon protein phosphorylation. Small regulatory molecules may also affect the strength of the association

Table 3. Sequence motifs in selected target proteins found or suggested to be involved in binding of 14-3-3 proteins. Full length sequences were searched for the 14-3-3 protein binding sites listed in Table 2 (T was allowed to substitute for S)

Protein (accession number)	Known 14-3-3 protein binding motif[a]	Putative 14-3-3 protein binding motif
A20 (P21580)	562-RSKSDP b	-
Glucocorticoid receptor (M10901)	Located to LBD (ligand binding domain) aa 485-777 c	42-SASSP 132-RSTSVP (no consensus site found in LBD)
Glycoprotein Iba (M22403)	612-DLLSTVSIRYSGHSL d; 623-GHSL e	378-KSTTEP
Glycoprotein Ibb (P13224)	189-RLSLTDP e	-
Inositol polyphosphate 5-phosphatase (P49442)	363-RSESEE f	-
Insulin receptor substrate-1 (S62539)	267-RSKpS g; 372-SRpSIP g; 639-SVpSAP g	-
NADH: nitrate reductase (S11868)	540-RTApSTP h	-
Phenylalanine hydroxylase (K03020)	-	270-RHGSKP
Protein kinase C (PKC) (L01087)	-	273-CQTKVANLC
Myosin II heavy chain specific protein kinase C (MHC-PKC) (M93393)	Located to CRD (cysteine rich domain) aa 1-139 i	68-CQTKVPLNC; 190-SPSSP; 635-RVRKVS
Plant H+-ATPase (P19456)	Located to C-terminal regulatory domain aa 850-947 j	693-RVKPSPTP k; 635-RSRSWS; 635-RELHTLK
Raf-1 (P04049)	257-RSTpSTP l; 619-RSApSEP l; 143-RKT m	176-CSTKVPTMC
Tryptophan hydroxylase (X52836)	-	258-RHSSDP
Tyrosine hydroxylase (X05290)	-	315-RHASSP n

[a] pS denotes a phosphorylated serine
[b] de Valck *et al.* 1997
[c] Wakui *et al.* 1997
[d] Du *et al.* 1996
[e] Andrews *et al.* 1998
[f] Campbell *et al.* 1997
[g] Ogihara *et al.* 1997
[h] Bachmann *et al.* 1996
[i] Matto-Yelin *et al.* 1997
[j] Jahn *et al.* 1997
[k] suggested by Aitken (1996)
[l] Muslin *et al.* 1996
[m] Clark *et al.* 1997
[n] suggested by Muslin *et al.* 1996

found in Raf-1 (Muslin *et al.* 1996), and the threonine in the RLSLpTDP motif found in the glycoprotein Ib-IX-V complex (Andrews *et al.* 1998) increases 14-3-3 protein binding affinity by at least an order of magnitude. This dependence suggests that protein kinases may control the affinity of 14-3-3 proteins for their targets. Most, if not all, 14-3-3-protein-binding motifs identified serve as substrates for serine/threonine kinases.

The kinases responsible for phosphorylation of 14-3-3 protein target sequences have been identified in a number of cases. For example, Raf-1 is phosphorylated *in vivo* at S43, S259, S499 and S621 (Kolch *et al.* 1993; McGrew *et al.* 1992; Morrison *et al.* 1993). S259 and S621 have both been implicated in 14-3-3 binding (Muslin *et al.* 1996). Phosphorylation of S621 seems to be required for Ras-1 functionality. Phosphorylation of S259 and S499 is apparently involved in protein kinase C-mediated activation of Raf (Kolch *et al.* 1993; Morrison *et al.* 1993; Sözeri *et al.* 1992).

14-3-3 interactions can also be regulated by means other than phosphorylation

It is intriguing that most 14-3-3 protein/protein interactions have been demonstrated by use of the yeast two-hybrid system (Table 1). If phosphorylation of specific sequences is a prerequisite for binding of 14-3-3 proteins to target proteins, then each of these heterologously expressed proteins must be phosphorylated at the correct site(s) for interaction to occur in the two-hybrid system. Since this prerequisite seems unlikely to be met in each case, it seems

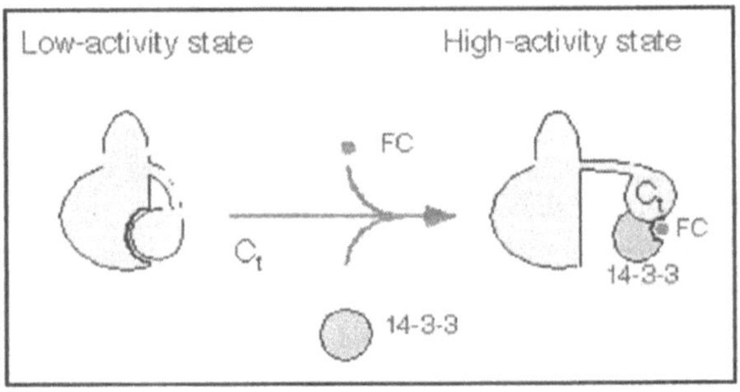

Fig. 4. Model for the regulation of plant plasma membrane H^+-ATPase by 14-3-3 proteins. The plasma membrane H^+-ATPase binds 14-3-3 protein. In the presence of the natural compound fusicoccin, the strength of this interaction increases several-fold. Neither the H^+-ATPase nor the 14-3-3 protein is able to bind fusicoccin alone. Rather the fusicoccin binding site is formed by both proteins. Binding of 14-3-3 protein to the C-terminal regulatory domain of the H^+-ATPase results in activation of the enzyme. It is not known whether the 14-3-3 protein binds to other sites of the H^+-ATPase

plausible that phosphorylation is not an all-or-nothing event, but rather serves to increase binding affinity of at least some target sequences. Furthermore, complex formation between 14-3-3 proteins and their targets do not always require phosphorylation; binding of 14-3-3 protein to the GHSL motif in the glycoprotein Ib-IX-V complex (Andrews *et al.* 1998) appears to be unaffected by phosphorylation. In addition, the N-terminal regulatory domain of c-Raf produced in *E. coli* binds 14-3-3 protein although it is not phosphorylated; in this case, the binding motif involved was not identified (Rommel *et al.* 1996).

Recently, it has been demonstrated that fusicoccin is able to stabilize the complex between 14-3-3 protein and the plant plasma membrane H^+-ATPase (Fig. 4) (Baunsgaard *et al.* 1998; Fullone *et al.* 1998; Jahn *et al.* 1997; Oecking *et al.* 1997). Fusicoccin induces tight binding of 14-3-3 protein to the plasma membrane H^+-ATPase, and as a consequence H^+-ATPase activity is increased, both *in vivo* and *in vitro* (Fig. 4) (Baunsgaard *et al.* 1998). Apparently no phosphorylation event is involved in this process since fusicoccin also induces the tight binding of recombinant 14-3-3 protein to a recombinant C-terminal domain of the H^+-ATPase (Fullone *et al.* 1998). Interestingly, the fusicoccin binding requires both 14-3-3 protein and H^+-ATPase (Baunsgaard *et al.* 1998). Neither protein binds fusicoccin alone. Therefore, it appears that 14-3-3 protein in concert with the target protein can generate a receptor for a regulatory molecule that affects the strength of the interaction between the two binding partners.

In the myosin II heavy-chain-specific protein kinase C, 14-3-3 protein binds to one of the CRDs (Matto-Yelin *et al.* 1997). The same could be true for other protein kinase C isoforms (Meller *et al.* 1996). Phorbol esters are natural compounds like fusicoccin which bind to and activate protein kinase C. Phorbol esters bind to at least one region in this protein, the CRD (residues 8–38) (Kazanietz *et al.* 1995a). In the presence of phosphatidylserine, binding of phorbol ester to this domain is further enhanced (Kazanietz *et al.* 1995b). It is not known whether phorbol esters affect the strength of the binding of the 14-3-3 protein to protein kinase C, but this remains an intriguing possibility.

Raf-1 is activated by Ras in a complicated fashion, which is still not understood, but it involves 14-3-3 proteins and translocation of Raf-1 to the plasma membrane. The N-terminal regulatory domain of Raf-1 contains two Ras-binding sites. The first is defined by residues 55–131; the second is located in the CRD (residues 139-184) (Brtva *et al.* 1995; Hu *et al.* 1995). This domain of Raf is homologous to CRDs found in protein kinase C family members. It appears that Ras first binds to Raf-1-(55–131) and that this binding unmasks the binding site in the Raf1-CRD which is normally cryptic (Drugan *et al.* 1996). In addition to other sites in the protein, the CRD has been identified as a site of interaction with 14-3-3 (Michaud *et al.* 1995). Mutation of Cys-165 and Cys-168 in Raf-1 CRD abolishes binding of both Ras and 14-3-3 (Michaud *et al.* 1995; Radziwill *et al.* 1996), whereas mutation of residues 143–145 disrupts binding of 14-3-3, but not Ras, to Raf-1-CRD (Clark *et al.* 1997). The latter mutant has increased transforming activity, suggesting that 14-3-3 binding to CRD inhibits Raf-1 activity (Clark *et al.* 1997). A lipid fraction derived from plasma membranes is able to activate Raf-1 *in vitro* (Dent and Sturgill 1994) and, interestingly, the CRD domain of Raf-1 also contains a binding site for phosphatidylserine (Ghosh *et al.* 1994). It is therefore possible that exposure of the CRD helps in guiding Raf-1 to the plasma membrane (Ghosh *et al.* 1994). Once in the membrane, phosphatidylserine interacts with 14-3-3 and Raf-1. It remains to be shown, however, whether phosphatidylserine in this system stabilizes the binding of 14-3-3 protein to Raf-1.

In conclusion, a picture is emerging with 14-3-3 protein as a general modulator of enzyme activity. Whereas calmodulin binds to and regulates proteins in a calcium-dependent manner, the action of 14-3-3 protein is dependent upon protein phosphorylation and/or the binding of small hydrophobic non-proteinaneous molecules. 14-3-3 Proteins might have different functions in different cells at different times of development. However, it appears that specificity might be generated by the target proteins present at a given time in that cell, and not by the 14-3-3 proteins themselves.

Acknowledgements. Work in the authors' laboratory was supported by the EU Biotechnology Programme, the Danish Veterinary and Agricultural Research Council and the Danish Biotech III Programme. We thank Thomas Hugh Roberts for critically reading the manuscript.

References

Acs, P., Szallasi, Z., Kazanietz, M.G., Blumberg, P.M. (1995) Differential activation of PKC isozymes by 14-3-3 zeta protein. *Biochem. Biophys. Res. Commun.* **216**:103–109

Aitken, A. (1996) 14-3-3 and its possible role in co-ordinating multiple signalling pathways. *Trends Cell. Biol.* **6**:341–347

Aitken, A., Howell, S., Jones, D., Madrazo, J., Martin, H., Patel, Y., Robinson, K. (1995) Post-translationally modified 14-3-3 isoforms and inhibition of protein kinase C. *Mol. Cell. Biochem.* **149/150**:41–49

Alam, R., Hachiya, N., Sakaguchi, M., Kawabata, S., Iwanaga, S., Kitajima, M., Mihara, K., Omura, T. (1994) cDNA cloning and characterization of mitochondrial import stimulation factor (MSF) purified from rat liver cytosol. *J. Biochem.* **116**:416–425

Andrews, R.K., Harris, S.J., McNally, T., Berndt, M.C. (1998) Binding of purified 14-3-3 z signaling protein to discrete amino acid sequences within the cytoplasmic domain of the platelet membrane glycoprotein Ib-IX-V complex. *Biochemistry* **37**:638–647

Bachmann, M., Huber, J.L., Liao, P.C., Gage, D.A., Huber, S.C. (1996) The inhibitor protein of phosphorylated nitrate reductase from spinach (*Spinacia oleracea*) leaves is a 14-3-3 protein. *FEBS Lett.* **387**:127–131

Banik, U., Wang, G.A., Wagner, P.D., Kaufman, S. (1997) Interaction of phosphorylated tryptophan hydroxylase with 14-3-3 proteins. *J. Biol. Chem.* **272**:26219–26225

Baunsgaard, L., Fuglsang, A.T., Jahn, T., Korthout, H.A.A.J., de Boer, A.H., Palmgren, M.G. (1998) The 14-3-3 protein associates with the plasma membrane H^+-ATPase to generate a fusicoccin binding complex and a fusicoccin responsive system. *Plant J.* **13**:661–671

de Boer, B. (1997) Fusicoccin – a key to multiple 14-3-3 locks? *Trends Plant. Sci.* **2**:60–66

Braselmann, S., McCormick, F. (1995) Bcr and Raf form a complex *in vivo* via 14-3-3 proteins. *EMBO J.* **14**:4839–4848

Brockhaus, K., Plaza, S., Pintel, D.J., Rommelaere, J., Salomé, N. (1996) Nonstructural proteins NS2 of minute virus of mice associate *in vivo* with 14-3-3 protein family members. *J. Virol.* **70**:7527–7534

Brtva, T.R., Drugan, J.K., Ghosh, S., Terrell, R.S., Campbell-Burk, S., Bell, R.M., Der, C.J. (1995) Two distinct Raf domains mediate interaction with Ras. *J. Biol. Chem.* **270**:9809–9812

Campbell, J.K., Gurung, R., Romero, S., Speed, C.J., Andrews, R.K., Berndt, M.C., Mitchell, C.A. (1997) Activation of the 43 kDa inositol polyphosphate 5-phosphataseby 14-3-3 zeta. *Biochemistry* **36**:15363-15370

Chang, H.C., Rubin, G.M. (1997) 14-3-3 epsilon positively regulates Ras-mediated signaling in Drosophila. *Genes Dev.* **11**:1132–1139

Clark, G.J., Drugan, J.K., Rossman, K.L., Carpenter, J.W., Rogers-Graham, K., Fu, H., Der, C.J., Campbell, S.L. (1997) 14-3-3z negatively regulates raf-1activity by interactions with the Raf-1 cysteine-rich domain. *J. Biol. Chem.* **272**:20990–20993

Conklin, D.S., Galaktionov, K., Beach, D. (1995) 14-3-3 proteins associate with cdc25 phosphatases. *Proc. Natl. Acad. Sci. USA* **92**:7892–7896

Craparo, A., Freund, R., Gustafson, T.A. (1997) 14-3-3 (epsilon) interacts with the insulin-like growth factor I receptor and insulin receptor substrate I in a phosphoserine-dependent manner. *J. Biol. Chem.* **272**:11663–11669

Cullere, X., Rose, P., Thathamangalam, U., Chatterjee, A., Mullane, K.P., Pallas, D.C., Benjamin, T.L., Roberts, T.M., Schaffhausen, B.S. (1998) Serine 257 phosphorylation regulates association of polyoma virus middle T antigen with 14-3-3 proteins. *J. Virol.* **72**:558–563

Daugherty, C.J., Rooney, M.F., Miller, P.W., Ferl, R.J. (1996) Molecular organization and tissue-specific expression of an *Arabidopsis* 14-3-3 gene. *Plant Cell* **8**:1239–1248

Dellambra, E., Patrone, M., Sparatore, B., Negri, A., Ceciliani, F., Bondanza, S., Molina, F., Cancedda, F.D., De Luca, M. (1995) Stratifin, a keratinocyte specific 14-3-3 protein, harbors a pleckstrin homology (PH) domain and enhances protein kinase C activity. *J. Cell Sci.* **108**:3569–3579

Dent, P., Sturgill, T.W. (1994) Activation of (His)6-Raf-1 *in vitro* by partially purified plasma membranes from v-Ras-transformed and serum-stimulated fibroblasts. *Proc. Natl. Acad. Sci. USA* **91**:9544–9548

Drugan, J.K., Khosravi-Far, R., White, M.A., Der, C.J., Sung, Y.J., Hwang, Y.W., Campbell, S.L. (1996) Ras interaction with two distinct binding domains inRaf-1 may be required for Ras transformation. *J. Biol. Chem.* **271**:233–237

Du, X., Fox, J.E., Pei, S. (1996) Identification of a binding sequence for the 14-3-3 protein within the cytoplasmic domain of the adhesion receptor, platelet glycoprotein Ib alpha. *J. Biol. Chem.* **271**:7362–7367

Fantl, W.J., Muslin, A.J., Kikuchi, A., Martin, J.A., MacNicol, A.M., Gross, R.W., Williams, L.T. (1994) Activation of Raf-1 by 14-3-3 proteins. *Nature* **371**:612–614

Ferl, R.J. (1996) 14-3-3 proteins and signal transduction. *Annu. Rev. Plant Physiol.* **47**:49–73

Freed, E., Symons, M., Macdonald, S.G., McCormick, F., Ruggieri, R. (1994) Binding of 14-3-3 proteins to the protein kinase raf and effects on its acivation. *Science* **265**:1713–1716

Fu, H., Coburn, J., Collier, J.R. (1993) The eukaryotic host factor that activates exoenzyme S of *Pseudomonas aeruginosa* is a member of the 14-3-3 protein family. *Proc. Natl. Acad. Sci. USA* **90**:2320–2324

Fullone, M.R., Visconti, S., Marra, M., Fogliano, V., Aducci, P. (1998) Fusicoccin effect on the *in vitro* interaction between plant 14-3-3 proteins and plasma membrane H+-ATPase. *J. Biol. Chem.* **273**:7698–7702

Furlanetto, R.W., Dey, B.R., Lopaczynski, W., Nissley, S.P. (1997) 14-3-3 proteins interact with the insulin-like growth factor receptor but not the insulinreceptor1. *Biochem. J.* **327**:765–771

Furukawa, Y., Ikuta, N., Omata, S., Yamauchi, T., Isobe, T., Ichimura, T. (1993) Demonstration of the phosphorylation-dependent interaction of tryptophan hydroxylase with the 14-3-3 protein. *Biochem. Biophys. Res. Commun.* **194**:144–149

Ghosh, S., Xie, W.Q., Quest, A.F., Mabrouk, G.M., Strum, J.C., Bell, R.M. (1994) The cysteine-rich region of raf-1 kinase contains zinc, translocates to liposomes, and is adjacent to a segment that binds GTP-ras. *J. Biol. Chem.* **269**:10000–10007

van Heusden, G.P.H., Griffiths, D.J., Ford, J.C., Chin, A., Woeng, T.F., Schrader, P.A., Carr, A.M., Steensma, H.Y. (1995) The 14-3-3 proteins encoded by the BMH1 and BMH2 genes are essential in the yeast *Saccharomyces cerevisiae* and can be replaced by a plant homologue. *Eur. J. Biochem.* **229**:45–53

van Heusden, G.P.H., Van der Zanden, A.L., Ferl, R.J., Steensma, H.Y. (1996) Four *Arabidopsis thaliana* 14-3-3 protein isoforms can complement the lethal yeast bmh1 bmh2 double disruption. *FEBS Lett.* **391**:252–256

Honda, R., Ohba, Y., Yasuda, H. (1997) 14-3-3z protein binds to the carboxyl half of mouse wee1 kinase. *Biochem. Biophys. Res. Commun.* **230**:262–265

Hsu, S.Y., Kaipia, A., Zhu, L., Hsueh, A.J. (1997) Interference of BAD (Bcl-xL/Bcl-2-associated death promoter)-induced apoptosis in mammalian cells by 14-3-3 isoforms and P11. *Mol. Endocrinol.* **11**:1858–1867

Hu, C.D., Kariya, K., Tamada, M., Akasaka, K., Shirouzu, M., Yokoyama, S., Kataoka, T. (1995) Cysteine-rich region of Raf-1 interacts with activator domain of post-translationally modified Ha-Ras. *J. Biol. Chem.* **270**:30274–30277

Ichimura, T., Isobe, T., Okuyama, T., Takahashi, N., Araki, K., Kuwano, R., Takahasi, Y. (1988) Molecular cloning of cDNA coding for brain-specific 14-3-3 protein, a protein kinase-dependent activator of tyrosine and tryptophan hydroxylases. *Proc. Natl. Acad. Sci. USA* **85**:7084–7088

Ichimura, T., Uchiyama, J., Kunihiro, O., Ito, M., Horigome, T., Omata, S., Shinkai, F., Kaji, H., Isobe, T. (1995) Identification of the site of interaction of the 14-3-3 protein with phosphorylated tryptophan hydroxylase. *J. Biol. Chem.* **270**:28515–28518

Ichimura, T., Ito, M., Itagaki, C., Takahashi, M., Horigome, T., Omata, S., Ohno, S., Isobe, T. (1997) The 14-3-3 protein binds its target proteins with a common site located towards the C-terminus. *FEBS Lett.* **413**:273–276

Irie, K., Gotoh, Y., Yashar, B.M., Errede, B., Nishida, E., Matsumoto, K. (1994) Stimulatory effects of yeast and mammalian 14-3-3 proteins on the Raf protein kinase. *Science* **265**:1716–1719

Isobe, T., Hiyane, Y., Ichimura, T., Okuyama, T., Takahashi, N., Nakajo, S., Nakaya, K. (1992) Activation of protein kinase C by the 14-3-3 proteins homologous with Exo1 protein that stimulates calcium-dependent exocytosis. *FEBS Lett.* **308**:121–124

Jahn, T., Fuglsang, A.T., Olsson, A., Brüntrup, I.M., Collinge, D.B., Volkmann, D., Sommarin, M., Palmgren, M.G., Larsson, C. (1997) The 14-3-3 protein interacts directly with the C-terminal region of the plant plasma membrane H+-ATPase. *Plant Cell* **9**:1805–1814

Jones, D.H., Ley, S., Aitken, A. (1995) Isoforms of 14-3-3 protein can form homo- and heterodimers *in vivo* and *in vitro*: implications for function as adapter proteins. *FEBS Lett.* **368**:5558

Kazanietz, M.G., Wang, S., Milne, G.W., Lewin, N.E., Liu, H.L., Blumberg, P.M. (1995a) Residues in the second cysteine-rich region of protein kinase C delta relevant to phorbol ester binding as revealed by site-directed mutagenesis. *J. Biol. Chem.* **270**:21852–21859

Kazanietz, M.G., Barchi, J.J. Jr., Omichinski, J.G., Blumberg, P.M. (1995b) Low-affinity binding of phorbol esters to protein kinase C and its recombinant cysteine-rich region in the absence of phospholipids. *J. Biol. Chem.* **270**:14679–14684

Knetsch, M.L.W., van Heusden, G.P.H., Ennis, H.L., Shaw, D.R., Epskamp, S.J.P., Snaar-Jagalska, B.E. (1997) Isolation of a *Dictyostelium discoideum* 14-3-3 homologue. *Biochim. Biophys. Acta* **1357**:243–248

Kockel, L., Vorbruggen, G., Jackle, H., Mlodzik, M., Bohmann, D. (1997) Requirement for *Drosophila* 14-3-3 zeta in Raf-dependent photoreceptor development. *Genes Dev.* **11**:1140–1147

Kolch, W., Heidecker, G., Kochs, G., Hummel, R., Vahidi, H., Mischak, H., Finkenzeller, G., Marme, D., Rapp, U.R. (1993) Protein kinase C alpha activates RAF-1 by direct phosphorylation. *Nature* **364**:249–252

Kosaki, A., Yamada, K., Suga, J., Otaka, A., Kuzuya, H. (1998) 14-3-3b protein associates with insulin receptor substrate 1 and decreases insulin-stimulated phosphatidylinositol 3'-kinase activity in 3T3L1 adipocytes. *J. Biol. Chem.* **273**:940–944

Kumagai, A., Yakowec, P.S., Dunphy, W.G. (1998) 14-3-3 proteins act as negative regulators of the mitotic inducer cdc25 in *Xenopus* egg extracts. *Mol. Biol. Cell.* **9**:345–354

Li, S., Sedivy, J.M. (1993) Raf-1 protein kinase activates the NF-kappa B transcription factor by dissociating the cytoplasmic NF-kappa B-I kappaB complex. *Proc. Natl. Acad. Sci. USA* **90**:9247–9251

Li, S., Janosch, P., Tanji, M., Rosenfeld, G.C., Waymire, J.C., Mischak, H., Kolch, W., Sedivy, J.M. (1995) Regulation of Raf-1 kinase activity by the 14-3-3 family of proteins. *EMBO J.* **14**:685–696

Li, W., Skoulakis, E.M., Davis, R.L., Perrimon, N. (1997) The *Drosophila* 14-3-3protein Leonardo enhances Torso signaling through D-Raf in a Ras 1-dependent manner. *Development* **124**:4163–4171

Liao, J., Omary, M.B. (1996) 14-3-3 proteins associate with phosphorylated simple epithelial keratins during cell cycle progression and act as a solubility cofactor. *J. Cell. Biol.* **133**:345–357

Liu, D., Bienkowska, J., Petosa, C., Collier, R.J., Fu, H., Liddington, R. (1995) Crystal structure of the zeta isoform of the 14-3-3 protein. *Nature* **376**:191–194

Liu, Y.C., Elly, C., Yoshida, H., Bonnefoy-Berard, N., Altman, A. (1996) Activation-modulated association of 14-3-3 proteins with Cbl in T cells. *J. Biol. Chem.* **271**:14591–14595

Liu, Y.C., Liu, Y., Elly, C., Yoshida, H., Lipkowitz, S., Altman, A. (1997) Serine phosphorylation of Cbl induced by phorbol ester enhances its association with 14-3-3 proteins in T cells via a novel serine-rich 14-3-3-binding motif. *J. Biol. Chem.* **272**:9979–9985

Lu, G., de Vetten, N.C., Sehnke, P.C., Isobe, T., Ichimura, T., Fu, H., van Heusden, G.P.H., Ferl, R.J. (1994) A single *Arabidopsis* GF14 isoform possesses biochemical characteristics of diverse 14-3-3 homologues. *Plant Mol. Biol.* **25**:659–667

Matto-Yelin, M., Aitken, A., Ravid, S. (1997) 14-3-3 inhibits the *Dictyostelium* myosin II heavy-chain-specific protein kinase C activity by a direct interaction: identification of the 14-3-3 binding domain. *Mol. Biol. Cell.* **8**:1889–1899

McGrew, B.R., Nichols, D.W., Stanton, V.P. Jr., Cai, H., Whorf, R.C., Patel, V., Cooper, G.M., Laudano, A.P. (1992) Phosphorylation occurs in the amino terminus of the Raf-1 protein. *Oncogene* **7**:33–42

Meller, N., Liu, Y.C., Collins, T.L., Bonnefoy-Bérard, N., Baier, G., Isakov, N., Altman, A. (1996) Direct interaction between protein kinase C theta (PKC theta) and 14-3-3 tau in T cells: 14-3-3 overexpression results in inhibition of PKCq translocation and function. *Mol. Cell. Biol.* **16**:5782–5791

Meyer, C., Waldkötter, K., Sprenger, A., Schlösser, U.G., Luther, M., Weiler, E.W. (1993) Survey of the taxonomic and tissue distribution of microsomal binding sites for the non-host selective fungal phytotoxin, fusicoccin. *Z. Naturforsch.* **48c**:595–602

Michaud, N.R., Fabian, J.R., Mathes, K.D., Morrison, D.K. (1995) 14-3-3 is not essential for Raf-1 function: identification of Raf-1 proteins that are biologically activated in a 14-3-3 and Ras-independent manner. *Mol. Cell. Biol.* **15**:3390–3397

Moore, B.W., Perez, V.J. (1967) Specific acidic proteins of the nervous system. In: Carlson, F.D. (ed) *Physiological and biochemical aspects of nervous integration.* Prentice-Hall, Englewood Cliff, N.J., pp. 343–359

Moorhead, G., Douglas, P., Morrice, N., Scarabel, M., Aitken, A., MacKintosh, C. (1996) Phosphorylated nitrate reductase from spinach leaves is inhibited by 14-3-3 proteins and activated by fusicoccin. *Curr. Biol.* **6**:1104–1113

Morrison, D.K., Heidecker, G., Rapp, U.R., Copeland, T.D. (1993) Identification of the major phosphorylation sites of the Raf-1 kinase. *J. Biol. Chem.* **268**:17309–17316

Muslin, A.J., Tanner, J.W., Alen, P.M., Shaw, A.S. (1996) Interaction of 14-3-3 with signaling proteins is mediated by the recognition of phosphoserine. *Cell* **84**:889–897

Oecking, C., Piotrowski, M., Hagemeier, J., Hagemann, K. (1997) Topology and target interaction of the fusicoccin-binding 14-3-3 homologs of *Commelina communis. Plant J.* **12**:441–453

Ogihara, T., Isobe, T., Ichimura, T., Taoka, M., Funaki, M., Sakoda, H., Onishi, Y., Inukai, K., Anai, M., Fukushima, Y., Kikuchi, M., Yazaki, Y., Oka, Y., Asano, T. (1997) 14-3-3 protein binds to insulin receptor substrate-1, one of the binding sites of which is in the phosphotyrosine binding domain. *J. Biol. Chem.* **272**:25267–25274

Papin, C., Denouel, A., Calothy, G., Eychène, A. (1996) Identification of signalling proteins interacting with B-Raf in the yeast two-hybrid system. *Oncogene* **12**:2213–2221

Radziwill, G., Steinhusen, U., Aitken, A., Moelling, K. (1996) Inhibition of Raf/MAPK signaling in *Xenopus* oocyte extracts by Raf-1-specific peptides. *Biochem. Biophys. Res. Commun.* **227**:20–26

Reuther, G.W., Fu, H., Cripe, L.D., Collier, R.J., Pendergast, A.M. (1994) Association of the protein kinases c-Bcr and Bcr-Abl with proteins of the 14-3-3 family. *Science* **266**:129–133

Roberts, R.L., Mosch, H.U., Fink, G.R. (1997) 14-3-3 proteins are essential for RAS/MAPK cascade signaling during pseudo hyphal development in *S. cerevisiae. Cell* **89**:1055–1065

Robertson, H., Langdon, W.Y., Thien, C.B., Bowtell, D.D. (1997) A c-Cbl yeast two-hybrid screen reveals interactions with 14-3-3 isoforms and cytoskeletal components. *Biochem. Biophys. Res. Commun.* **240**:46–50
Robinson, K., Jones, D., Patel, Y., Martin, H., Madrazo, J., Martin, S., Howell, S., Elmore, M., Finnen, M.J., Aitken, A. (1994) Mechanism of inhibition of protein kinase C by 14-3-3 isoforms: 14-3-3 isoforms do not have phospholipaseA2 activity. *Biochem. J.* **299**:853–861
Rommel, C., Radiziwill, G., Lovric, J., Noeldeke, J., Heinicke, T., Jones, D., Aitken, A., Moelling, K. (1996) Activated Ras displaces 14-3-3 protein from the amino terminus of c-Raf-1. *Oncogene* **12**:609–619
Rommel, C., Radziwill, G., Moelling, K., Hafen, E. (1997) Negative regulation of Raf activity by binding of 14-3-3 to the amino terminus of Raf in vivo. *Mech. Dev.* **64**:95–104
Sehnke, P.C., Ferl, R.J. (1996) Plant metabolism: enzyme regulation by 14-3-3proteins. *Curr. Biol.* **6**:1403–1405
Sladeczek, F., Camonis, J.H., Burnol, A.F., Le Bouffant, F. (1997) The Cdk-like protein PCTAIRE-1 from mouse brain associates with p11 and 14-3-3 proteins. *Mol. Gen. Genet.* **254**:577
Sözeri, O., Vollmer, K., Liyanage, M., Frith, D., Kour, G., Mark, G.E., Stabel, S. (1992) Activation of the c-Raf protein kinase by protein kinase C phosphorylation. *Oncogene* **7**:2259–2262
Suen, K.L., Bustelo, X.R., Barbacid, M. (1995) Lack of evidence for the activation of the Ras/Raf mitogenic pathway by 14-3-3 proteins in mammalian cells. *Oncogene* **11**:825–831
Tanji, M., Horwitz, R., Rosenfeld, G., Waymire, J.C. (1994) Activation of protein kinase C by purified brain 14-3-3:comparison with tyrosine hydroxylase activation. *J. Neurochem.* **63**:1908–1916
Toker, A., Sellers, L.A., Amess, B., Patel, Y., Harris, A., Aitken, A. (1992) Multiple isoforms of a protein kinase inhibitor (KCIP-1/14-3-3) from sheep brain: amino acid sequences of phosphorylated forms. *Eur. J. Biochem.* **206**:453–461
de Valck, D., Heyninck, K., Van Criekinge, W., Vandenabeele, P., Fiers, W., Beyaert, R. (1997) A20 inhibits NF-kappaB activation independently of binding to14-3-3 proteins. *Biochem. Biophys. Res. Commun.* **238**:590–594
de Vetten, N.C., Lu, G., Ferl, R.J. (1992) A maize protein associated with the G-box binding complex has homology to brain regulatory proteins. *Plant Cell* **4**:1295–1307
Vincenz, C., Dixit, V.M. (1996) 14-3-3 proteins associate with A20 in an isoform-specific manner and function both as chaperone and adapter molecules. *J. Biol. Chem.* **271**:20029–20034
Wakui, H., Wright, A.P.H., Gustafsson, J.-Å., Zilliacus, J. (1997) Interaction of the ligand-activated glucocorticoid receptor with the 14-3-3h protein. *J. Biol. Chem.* **272**:8153–8156

Wang, W., Shakes, D.C. (1996) Molecular evolution of the 14-3-3 protein family. *J. Mol. Evol.* **43**:384–398

Wang, W., Shakes, D.C. (1997) Expression patterns and transcript processing of fft-1 and fft-2, two *C. elegans* 14-3-3 homologues. *J. Mol. Biol.* **268**:619–630

Watanabe, M., Isobe, T., Okuyama, T., Ichimura, T., Kuwano, R., Takahashi, Y., Kondo, H. (1991) Molecular cloning of cDNA to rat 14-3-3 eta chain polypeptide and the neuronal expression of the mRNA in the central nervous system. *Mol. Brain. Res.* **10**:151–158

Watanabe, M., Isobe, T., Ichimura, T., Kuwano, R., Takahashi, Y., Kondo, H. (1993a) Molecular cloning of rat cDNAs for beta and gamma subtypes of 14-3-3 protein and developmental change in expression of their mRNAs in the nervous system. *Mol. Brain Res.* **17**:135–146

Watanabe, M., Isobe, T., Ichimura, T., Kuwano, R., Takahashi, Y., Kondo, H. (1993b) Developmental regulation of neuronal expression for the eta subtype of the14-3-3 protein, a putative regulatory protein for protein kinase C. *Dev. Brain Res.* **73**:225–235

Watanabe, M., Isobe, T., Ichimura, T., Kuwano, R., Takahashi, Y., Kondo, H., Inoue, Y. (1994) Molecular cloning of rat cDNAs for the zeta and theta subtypes of 14-3-3 protein and differential distribution of their mRNAs in brain. *Mol. Brain Res.* **25**:113–121

Wu, K., Rooney, M.F., Ferl, R.J. (1997a) The *Arabidopsis* 14-3-3 multigene family. *Plant Physiol.* **114**:1421–1431

Wu, K., Lu, G., Sehnke, P., Ferl, R.J. (1997b) The heterologous interactions among plant 14-3-3 proteins and identification of regions that are important for dimerization. *Arch. Biochem. Biophys.* **339**:2–8

Xiao, B., Smerdon, S.J., Jones, D.H., Dodson, G.G., Soneji, Y., Aitken, A., Gamblin, S.J. (1995) Structure of a 14-3-3 protein and implications for coordination of multiple signalling pathways. *Nature* **376**:188–191

Xing, H., Kornfeld, K., Muslin, A.J. (1997) The protein kinase KSR interacts with 14-3-3 protein and Raf. *Curr. Biol.* **7**:294–300

Yaffe, M.B., Rittinger, K., Volinia, S., Caron, P.R., Aitken, A., Leffers, H., Gamblin, S.J., Smerdon, S.J., Cantley, L.C. (1997) The structural basis for14-3-3:phosphopeptide binding specificity. *Cell* **91**:961–971

Zha, J., Harada, H., Yang, E., Jockel, J., Korsmeyer, S.J. (1996) Serine phosphorylation of death agonist BAD in response to survival factor results in binding to14-3-3 not BCL-X. *Cell 87*:619–628

Zhang, H., Wang, J., Nickel, U., Allen, R.D., Goodman, H.M. (1997) Cloning and expression of an *Arabidopsis* gene encoding a putative peroxisomal ascorbate peroxidase. *Plant Mol. Biol.* **34**:967–971

Zhang, L., Wang, H., Liu, D., Liddington, R., Fu, H. (1997) Raf-1 kinase and exoenzyme S interact with 14-3-3z through a common site involving lysine 49. *J. Biol. Chem.* **272**:13717–13724

Zhang, S.H., Kobayashi, R., Graves, P.R., Piwnica-Worms, H., Tonks, N.K. (1997) Serine phosphorylation-dependent association of the band 4.1-related protein-tyrosine phosphatase PTPH1 with 14-3-3b protein. *J. Biol. Chem.* 272:27281–27287

Enzymes of urate synthesis and catabolism in the Gecarcinid land crab *Gecarcoidea natalis*

Stuart M. Linton · Peter Greenaway

School of Biological Science, University of New South Wales, Sydney, Australia 2052

Received: 31 March 1998 / Accepted: 13 May 1998 / Published: 17 June 1998

Abstract. This study investigated the sites of urate synthesis and catabolism in the gecarcinid land crab *Gecarcoidea natalis* by assaying spongy connective tissue, midgut gland, muscle and gill for xanthine oxidoreductase, the last enzyme involved in urate synthesis, and uricase and urease, the first and last enzymes involved in urate catabolism. The spongy connective tissue and midgut gland of the *G. natalis* contained activities of xanthine oxidoreductase and were considered to be sites of urate synthesis. The midgut gland had a high activity of xanthine oxidoreductase [(58.87±4.6 (SE) nmol urate produced g^{-1} wet wt. tissue min^{-1}], 2.7 times the xanthine oxidoreductase activity contained within the spongy connective tissue, and was thought to be the main site of urate synthesis. Xanthine dehydrogenase (EC 1.1.1.204) was the only form of xanthine oxidoreductase detected within the tissues. Its presence means that the cost of synthesising urate *de novo* is relatively small (between 1 and 3 ATP). Uricase (EC 1.7.3.3) and urease (EC 3.5.1.5) activities were present in the tissues of *G. natalis*. Spongy connective tissue contained the highest activities of uricase [48.44±4.29 (SE) nmol urate consumed g^{-1} wet wt. tissue min^{-1}] while the highest activities of urease [365.31±37.21 (SE) nmol urate consumed g^{-1} wet wt tissue min^{-1}] were contained within the gills. From this evidence it is clear that *G. natalis* possesses the uricolytic pathway and hence the ability to catabolise urate, and urate catabolism is begun at the site of urate storage, the spongy connective tissue, and is completed at the gills. As the gills are the site of ammonia excretion in this species the ammonia produced from the catabolism of urate is probably excreted. The urate deposits within the body of *G. natalis* may be involved in temporary storage of nitrogenous wastes.

Key words. *Gecarcoidea natalis* · Land crab · Urate metabolism · Urease Uricase · Xanthine dehydrogenase

Correspondence to: S.M. Linton (e-mail s.linton@unsw.edu.au, Fax: +61-2-93851558)

Introduction

Urate deposits are present within the bodies of numerous terrestrial and aquatic crustaceans including terrestrial and aquatic decapods and isopods, and aquatic amphipods (for review of species see Greenaway 1991; O'Donnell and Wright 1995). These urate deposits, in both terrestrial [e.g. *Gecarcoidea natalis* (gecarcinid land crab)] and aquatic crustaceans [e.g. *Niphargus schellenbergi* (amphipod), *Asellus aquaticus* (isopod), *Callinectes sapidus* (decapod) and *Carcinus maenas* (decapod)], are intracellular, occurring in connective tissue cells as membrane-bound urate crystals of approximately 1 μm in diameter (Fischer 1926; Graf and Michaut 1975; Johnson 1980; Wägele 1992; Linton and Greenaway 1997a). In *G. natalis*, and probably other land crabs, the urate deposits are synthesised *de novo* from excess dietary nitrogen (Linton and Greenaway 1997b).

It is not known in which tissues urate is synthesised in gecarcinid crabs. There are two obvious possibilities: the spongy connective tissue where the urate is stored and the midgut gland, or hepatopancreas, where dietary protein is digested and assimilated (Icely and Nott 1992). Spongy connective tissue is unlikely to be the main site of urate synthesis since it contains relatively few organelles and hence is not very active metabolically (Linton and Greenaway 1997a). Synthetic activity is likely to be located in the midgut gland since this tissue is involved in urate synthesis in *Birgus latro* and xanthine oxidoreductase, the last enzyme involved in urate formation, is present in the midgut gland of numerous terrestrial and aquatic crustaceans (Greenaway and Morris 1989; Dykens 1991; Lallier and Walsh 1991).

Purine synthesis *de novo* occurs in evolutionarily diverse species such as the branchiopod *Artemia salina*, the anomuran land crab *B. latro* and the gecarcinid land crab *G. natalis* and it is becoming clear that crustaceans, generally, may possess this ability (Greenaway and Morris 1989; Liras *et al.* 1992; Linton and Greenaway 1997b). The enzymatic steps utilised by crustaceans to synthesise urate *de novo* are thought to be similar to those already described for insects and birds (Linton and Greenaway 1997b). Thus synthesis of urate first requires construction of the purine ring. This begins with 5'- phosphoribosyl 1- pyrophosphate and produces the nucleotide IMP by addition of the following precursors; one glycine (carbons C4, C5 and nitrogen N7), two glutamines (nitrogens N3 and N9), one aspartate (nitrogen N1), two N^{10}-formyl tetrahydrofolate derivatives (carbons C2 and C8) and carbon dioxide (carbon 9) (Stryer 1988; Campbell 1995). Urate is then derived from IMP in several steps. In the first, IMP is dephosphorylated to form the nucleoside inosine. The glycosidic bond of inosine is then cleaved by purine nucleoside phosphorylase to create hypoxanthine and ribose (Stryer 1988). The ribose could be regenerated to 5'- phosphoribosyl 1- pyrophosphate and used to synthesise a further purine molecule (Campbell 1995). This negates the cost of synthesising the sugar. The hypoxanthine is then oxidised to xanthine which in turn is oxidised to uric acid (Campbell 1995). These oxidation reactions are catalysed by xanthine oxidoreductase, the last enzyme in the urate *de novo* synthetic pathway (Campbell 1995).

Xanthine oxidoreductase has two forms, a dehydrogenase and an oxidase. Xanthine dehydrogenase (EC 1.1.1.204) uses NAD^+ as the electron acceptor and produces NADH (Campbell 1995). Xanthine oxidase (EC 1.1.3.22) uses molecular oxygen (O_2) as the electron acceptor and produces either superoxide (O_2^-) or hydrogen peroxide (H_2O_2) (Stryer 1988). These species are highly reactive and since they could potentially damage the cell (McCord 1985) would normally be catabolised by superoxide dismutase and catalase within the peroxisomes (Stryer 1988). Xanthine dehydrogenase can be converted to xanthine oxidase, either irreversibly by limited proteinolysis with proteinolytic enzymes such as trypsin or chymotrypsin, or reversibly by the oxidation of the thiol groups (Della Corte and Stirpe 1968, 1972). In all species studied, including vertebrates and insects, xanthine oxidoreductase *in vivo* occurs as the dehydrogenase (Battelli *et al.* 1972; reviewed Kooij 1994; Campbell 1995; Harrison 1995). Similarly xanthine dehydrogenase is also the main component of the xanthine oxidoreductase activity in the majority of the crustacean species studied so far (Dykens 1991; Lallier and Walsh 1991). *B. latro* is the only exception, where xanthine oxidoreductase has been reported to be present as the oxidase form (Greenaway and Morris 1989).

Despite the high initial metabolic cost of synthesising the purine ring, the cost of synthesising urate *de novo* is relatively small due to the presence of xanthine dehydrogenase. Synthesis of IMP requires nine ATP: three ATP for the synthesis of phosphoribosyl pyrophosphate (PPRP) from ribose, four ATP for the synthesis of IMP from PPRP and other precursors mentioned previously and two ATP for the formation of the two N^5-formyl tetrahydrofolate derivatives (Campbell 1995). If N^5, N^{10}-methylene tetrahydrofolate is used as a precursor for purine synthesis instead of N^5-formyl tetrahydrofolate then the total cost of synthesising IMP is seven ATP since the formation of N^5, N^{10}-methylene tetrahydrofolate requires no ATP (Stryer 1988). In the two last steps of urate synthesis, the oxidation of hypoxanthine to urate by xanthine dehydrogenase, two NADH are formed. When these two NADH are oxidised by the electron transport chain, six ATP are produced. Thus by summation the total cost of synthesising urate *de novo* is between one and three ATP.

The uricolytic pathway is present in the midgut gland of aquatic decapods *Menippe mercenaria*, *Callinectes sapidus* and *Carcinus maenas*, the stomatopod *Oratosquilla oratoria* and the midgut gland, intestine and body wall of the terrestrial isopod *Oniscus asellus* (Noguchi *et al.* 1979; Claybrook 1983; Dykens 1991; Lallier and Walsh 1991). The occurrence of uricolytic enzymes within the tissues of gecarcinid crabs has not been well studied. The midgut glands of the gecarcinid land crabs *C. guanhumi* and *G. lateralis* are reported not to contain any uricolytic enzymes (Lallier and Walsh 1991) but it is possible that the uricolytic enzymes as in the isopod *Oniscus asellus* are present in a tissue other than the midgut gland.

The uricolytic pathway by which urate is catabolised to ammonia has four steps (Stryer 1988). The first step, catalysed by uricase (EC 1.7.3.3), is the oxidation of urate to allantoin (Stryer 1988). Allantoin is hydrolysed to allantoic acid which in turn is hydrolysed to urea and glyoxylate (Stryer 1988). These reactions are respectively catalysed by allantoinase and allantoicase. Finally urease (EC 3.5.1.5) hydrolyses urea to ammonia and CO_2 (Stryer 1988).

This study examined the site of urate synthesis and catabolism in the gecarcinid land crab, *G. natalis*, by establishing the presence and activity of key enzymes involved in the pathways. This was achieved by assaying various tissues (midgut gland, muscle, branchiostegites and gill) for the last enzyme in the urate synthesis pathway, xanthine oxidoreductase, and the uricolytic enzymes uricase and urease. The presence and location of the uricolytic enzymes was used as a basis for suggestions as to the possible physiological roles of urate deposits in *G. natalis*.

Materials and methods

Collection and maintenance of animals

Gecarcoidea natalis (Pocock) of both sexes, weighing 200–300 g each, were collected from rainforest on Christmas Island, Indian Ocean and air freighted to Sydney. In the laboratory, they were maintained in large communal containers at 25°C and 80% humidity with a 12-h light, 12-h dark cycle. Crabs were supplied with tap water for drinking and fed fallen leaves of *Ficus macrophylla* and occasional soya beans and apples.

Homogenisation of tissues for xanthine oxidoreductase and uricase assays

Animals were chilled until torpid and killed by rapidly removing the heart and dissecting out the required tissues (midgut gland, branchiostegites, leg muscle and gill). The tissues were blotted, weighed and homogenised (50 strokes) using a ground glass homogeniser (Wheaton), in 4–6 volumes of ice-cold 25 mmol l^{-1} Tris buffer (pH=7.5) containing 1 mmol l^{-1} dithioerythritol (DTE), 6 mmol l^{-1} Na_2EDTA and 0.2 mmol l^{-1} phenylmethylsulphonyl fluoride (PMSF) (proteinase inhibitor). Prior to use, 2 µl of PMSF stock (0.1 mol l^{-1}) was added to each millilitre of buffer. The PMSF was dissolved in isopropanol at -20°C overnight. The homogenate was then centrifuged at 10,000 *g* for 10 min and the supernatant used for assays.

Tissue homogenisation for urease assays

Homogenates for urease assays were prepared as described above, except that the tissues were homogenised in 25 mmol l^{-1} potassium phosphate buffer at pH=7.5 containing 1 mmol l^{-1} DTE. Phosphate buffers was used for urease assays instead of Tris buffers because Tris was found to interfere with the Solarozano colour reaction of the Sigma urea kit (Cat. no. 640-A).

Xanthine oxidoreductase assays

Enzyme activities were assayed spectrophotometrically using a LKB Ultrospec II spectrophotometer and methacrylate (UV) cuvettes with a 1-cm light path.

Xanthine oxidase

To remove endogenous uric acid from the homogenates 120 µl of the homogenate was incubated with 1650 µl of assay buffer (100 mmol l^{-1} pH=7.5 Tris buffer with 1 mmol l^{-1} DTE) and 30 µl of uricase (18 U/ml) (Boehringer Mannheim Cat. no. 737 364) at 25°C for 10 min. After this period the xanthine oxidase reaction was initiated by adding the substrate (200 µl of 0.2 mmol l^{-1} xanthine). Xanthine oxidase activity was measured at 25°C by continuously monitoring the disappearance of xanthine at 270 nm for 10 min.

Xanthine dehydrogenase

Endogenous uric acid was removed from the homogenate as described for xanthine oxidase. The xanthine dehydrogenase reaction was started by adding 200 µl of 0.2 mmol l^{-1} xanthine and 100 µl of 12 mmol l^{-1} NAD^{+}. The reaction was then monitored continuously for 10 min by following the appearance of NADH at 340 nm.

Uricase assays

Activities of uricase were assayed by adding 40 µl of homogenate to 1760 µl of uricase assay buffer (100 mmol l^{-1} pH=8.5 Tris buffer with 1 mmol l^{-1} DTE), and 200 µl of 0.2 mmol l^{-1} uric acid and incubating at 25 C for 20 min. Uricase activity was monitored continuously by following the disappearance of uric acid at 293 nm.

Urease assays

Urease assays were conducted by adding 100 µl of homogenate to 100 µl of 10 mmol l^{-1} urea and 740 µl of urease assay buffer (25 mmol l^{-1} pH=7.5 phosphate buffer containing 1 mmol l^{-1} DTE) and incubating the mixture at 25°C for 20 min. Blanks without the urea substrate contained 100 µl of the homogenate and 840 µl of the urease assay buffer. At the end of the incubation period the reactions were stopped by adding 50 µl of 0.6 mmol l^{-1} $HClO_4$ to precipitate the enzyme. Addition of 10 µl of 2.5 mmol l^{-1} K_2CO_3 neutralised the suspension. Precipitated protein was pelleted by centrifuging the reaction mixture at 10,000 *g* for 20 min. Ammonia concentrations in the supernatant of experimental and blank tubes were then measured with a Sigma urea nitrogen kit (Cat. no. 640-A). The difference in the amounts of ammonia in the experimental and blank assay systems was the amount of ammonia produced in 20 min by the hydrolysis of urea by urease.

Statistics

Data are expressed as mean ± SE. Statistical comparisons between means were made using one way analysis of variance (ANOVA) and LSD a posteriori tests (SPSS for Windows 6.0).

Results

Urate synthesis enzyme (xanthine oxidoreductase)

Xanthine dehydrogenase was the only form of xanthine oxidoreductase detected within the tissues of *G. natalis* (Table 1). Xanthine dehydrogenase activities were present in the midgut gland and branchiostegites but not in the muscle and gill of any crab assayed (Table 1). Activity of xanthine dehydrogenase in the midgut gland was 2.7 times the xanthine dehydrogenase activity in the branchiostegite (Table 1).

Enzymes of the uricolytic pathway

Uricase

Uricase activities were present in the branchiostegites (spongy connective tissue), midgut gland and gill tissue of *G. natalis* (Table 1). In all animals assayed, the branchiostegites contained the highest activity of uricase (Table 1). Uricase activity was also evident in the midgut gland and gills (Table 1). The mean uricase activities in the midgut gland and gills were low, as only 11.1% of animals assayed exhibited uricase activity in their midgut glands and only 37.5% of crabs assayed contained uricase activity in their gills (Table 1).

Urease

Measurable levels of urease activity were found only in the midgut gland and gills of *G. natalis* (Table 1). The gills of all the animals assayed contained urease activity at a level 4.1 times the urease activity in the midgut gland (Table 1). The urease activity within the gills was also higher than the activities of xanthine oxidoreductase in the midgut gland and branchiostegites (6.2 and 16.5 times respectively) and the activity of uricase in the branchiostegites

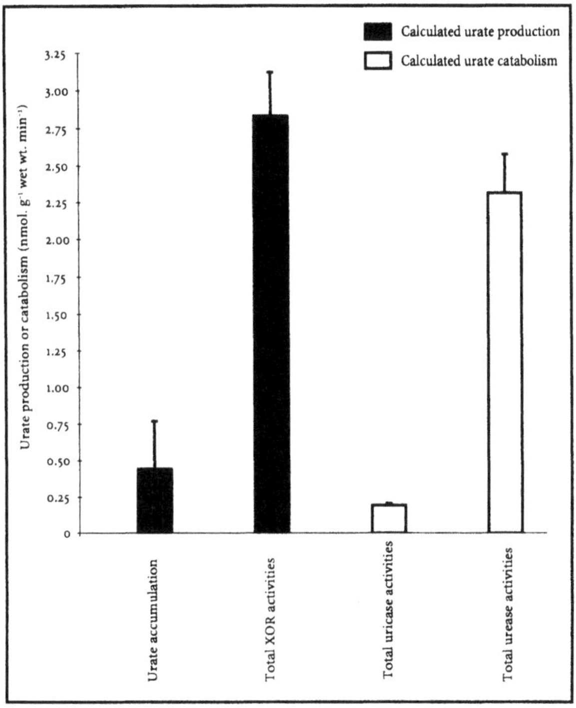

Fig. 1. Rates of urate production and catabolism in *Gecarcoidea natalis*. Rates are calculated from values for urate accumulation by animals maintained on a high nitrogen diet for 6 weeks (Linton and Greenaway 1997b) and from total enzyme activities of xanthine oxidoreductase (n=7), uricase (n=10) and urease (n=7) Rates of urate production or catabolism (mean ± SE) are expressed as nmol urate produced or consumed g^{-1} wet wt. crab min^{-1}

Table 1. Activities of synthetic (xanthine oxidoreductase) and catabolic (uricase and urease) enzymes of urate metabolism in the tissues of *Gecarcoidea natalis*. Xanthine dehydrogenase (*XDH*) and xanthine oxidase (*XO*) were the two forms of xanthine oxidoreductase that were assayed. Enzyme activities [mean ± SE (n)] are expressed as nmoles substrate produced (uric acid) or consumed (uric acid or urea) g^{-1} wet wt. tissue min^{-1}. Different *superscripts* within a *row* indicate that the means differed significantly (P<0.05). (*ND* Enzyme activity not detected)

Enzymes of synthesis:	**Midgut gland**	**Branchiostegite**	**Muscle**	**Gill**
XDH activity	58.87 ± 4.76^a (10)	22.10 ± 4.79^b (10)	ND (8)	ND (8)
% Animals with XDH	100	100	0	0
XO activity	ND (10)	ND (10)	ND (10)	ND (10)
Catabolic enzymes:				
Uricase activity	3.66 ± 3.66^b (9)	48.44 ± 4.29^a (8)	ND (8)	13.51 ± 8.47^b (8)
% Animals with uricase	11.1	100	–	37.5
Urease activity	88.28 ± 78.35^a (8)	ND (8)	ND (8)	365.31 ± 37.21^b (8)
% Animals with urease	75	–	–	100

(7.5 times) (Table 1, Fig. 1). Urease activity in the midgut gland was not detected in all of the animals assayed (Table 1), and the values were variable: 62.5% of the animals assayed exhibited low activities (5.4–22.1 nmol urea g^{-1} wet wt. tissue min^{-1}) and 12.5% of the crabs assayed had high activities (636.35 nmol urea g^{-1} wet wt. tissue min^{-1}).

Total enzyme activity

The total activities of xanthine oxidoreductase, uricase and urease within the various tissues and hence the animal were calculated as the product of the tissue enzyme activity (nmol substrate consumed g^{-1} wet tissue min^{-1}) and the total weight of that tissue divided by the wet weight of the animal. Using these activities, the rates of urate production or catabolism were calculated (Fig. 1).

The enzymes, xanthine dehydrogenase and uricase, within the branchiostegites are probably contained in spongy connective tissue cells since these are the main cell type present (Farrelly and Greenaway 1993). Spongy connective tissue is spread throughout the body (Linton and Greenaway 1997a) and hence total enzyme activities of xanthine oxidoreductase and urease calculated for the branchiostegites are probably underestimates of the total activity contained within the spongy connective tissue.

Rates of urate production and catabolism

The rate of production of urate calculated from the total xanthine dehydrogenase activity was 6.3 times the measured rate of urate accumulation in crabs maintained on a high-nitrogen diet for 6 weeks (Linton and Greenaway 1997b) (Fig. 1). The rate of urate oxidation and hence catabolism calculated from the total uricase activities in the branchiostegites amounted to 7% of the rate of urate production based on total xanthine dehydrogenase activity and 40% of the rate of urate accumulation in animals maintained on a high-nitrogen diet for 6 weeks (Fig. 1). Urease activity in the gills could hydrolyse the two urea molecules resulting from the catabolism of urate 12.8 times more rapidly than uricase could oxidise urate (Fig. 1).

Discussion

Xanthine oxidoreductase; final enzyme of urate synthesis

Sites of urate synthesis

The presence of xanthine dehydrogenase activity in the midgut gland and branchiostegite of *G. natalis* indicates that these tissues are sites of urate synthesis (Table 1). The midgut gland is probably the main site of urate synthesis, since it is a large organ (5.05±0.19% of wet wt.) (S. Linton unpublished data) and contains the highest activities of xanthine dehydrogenase (Table 1). On the basis of the location and level of activity of xanthine oxidoreductase, the majority of the urate is synthesised *de novo* in the midgut gland from excess dietary amino acids and then transported, probably via the haemolymph, to the spongy connective tissue cells where it is stored intracellularly as urate crystals (Linton and Greenaway 1997a,b). The midgut gland of the anomuran land crab *B. latro* is also thought to be the site of urate synthesis given that this tissue exhibits high activities of xanthine oxidoreductase (Greenaway and Morris 1989). The presence of xanthine oxidoreductase in the midgut glands of other terrestrial (e.g. *Cardisoma guanhumi*, *Gecarcinus lateralis*) and aquatic (e.g. *Menippe mercenaria*, *Callinectes sapidus*, *Carcinus maenas*) decapods suggests this may be the site of urate formation in decapods generally (Dykens 1991; Lallier and Walsh 1991).

Xanthine dehydrogenase activity was present in the spongy connective tissue of the branchiostegites and presumably other spongy connective tissue sites throughout the body. The paucity of organelles within the spongy connective tissue (Linton and Greenaway 1997a; Farrelly and Greenaway 1993) and the low activity of xanthine dehydrogenase within the branchiostegites (spongy connective tissue) suggest that this tissue may not be the primary site of urate *de novo* synthesis (Table 1).

Rates of urate synthesis

The rate of urate synthesis in *G. natalis* may be higher than in other gecarcinid land crabs (*C. guanhumi* and *G. lateralis*), known to accumulate urate within their bodies, since the xanthine dehydrogenase activity in the midgut gland of *G. natalis* was higher than that reported in the midgut gland of other gecarcinid crabs (Table 2) (Gifford 1968; Horne 1968; Wolcott and Wolcott 1984, 1987; Lallier and Walsh 1991). The anomuran land crab *B. latro* produces large amounts of uric acid for excretion of its nitrogenous wastes and it is not surprising that its midgut gland had the highest xanthine oxidoreductase activity of all the species studied (Table 2) (Greenaway and Morris 1989).

The total xanthine oxidoreductase activity within the body of *G. natalis* would potentially allow a rate of urate synthesis that is considerably higher than the rate of accumulation actually measured in this species when it was kept on a high-nitrogen diet (Fig. 1) (Linton and Greenaway 1997b). It is probable, therefore, that xanthine oxidoreductase is not the rate-limiting enzyme of urate synthesis. This is certainly the case for termites, where xanthine oxidoreductase activities are 4–11 times the activity of the rate limiting enzyme, purine nucleoside phosphorylase (Chappell and Slaytor 1993). *In vivo* substrate concentrations may also be lower than those used in the assays and hence the rate of urate production would be lower.

Xanthine dehydrogenase

In the tissues of *G. natalis*, xanthine dehydrogenase was the only form of xanthine oxidoreductase present (Table 2). This is consistent with the majority of other crustaceans studied in which the dehydrogenase form of xanthine oxidoreductase was also dominant (Table 2). Only in *B. latro* has xanthine oxidase been reported to be the major form of xanthine oxidoreductase (Table 2) (Greenaway and Morris 1989). This may however be an artefact, as a protease inhibitor was not included in the buffer used to prepare the homogenates of the midgut gland of *B. latro* (Greenaway and Morris 1989) and it is possible that xanthine dehydrogenase was converted to xanthine oxidase by limited damage with proteinolytic enzymes. Similarly xanthine dehydrogenase has been found *in vivo* in the tissues of all animals (vertebrates and invertebrates) that have xanthine oxidoreductase (Battelli *et al.* 1972; reviewed Kooij 1994; Campbell 1995; Harrison 1995).

Table 2. Xanthine oxidoreductase (*XOR*) activities in the midgut glands of terrestrial crabs. (*XO* Activities of xanthine oxidase, *XDH* Activities of xanthine dehydrogenase, *total XOR activity* XO activity+XDH activity.) Enzyme activities (mean ± SE) are expressed as nmol urate produced g^{-1} wet wt. tissue min^{-1}

Species	XO	XDH	Total XOR	Reference
Terrestrial crabs:				
Anomura				
Birgus latro	66.5±16.9 - 81.5±17.5	0	66.5±16.9 - 81.5±17.2	Greenaway and Morris (1989)
Brachyura				
Gecarcoidea natalis		58.9 ± 4.8	58.9 ± 4.8	This study
Cardisoma guanhumi	2.9 ± 1.3	20.6 ± 7.7	23.5	Lallier and Walsh (1991)
Gecarcinus lateralis	2.2 ± 1.1	1.7 ± 1.3	3.9	Lallier and Walsh (1991)

As mentioned in the Introduction the presence of xanthine dehydrogenase rather than xanthine oxidase in the tissues of *G. natalis* means that the total cost of synthesising urate *de novo* is relatively small, probably being between one and three molecules of ATP. Indeed at this price, synthesis of urate *de novo* is cheaper than synthesising urea (four ATP). Urate synthesis *de novo* has previously been considered to be energetically expensive (seven to nine ATP) since xanthine oxidase was initially thought to be the form of xanthine oxidoreductase involved in urate formation (Kooij 1994). The detoxification and excretion of nitrogenous wastes as uric acid is therefore not as energetically disadvantageous as has been commonly assumed.

Enzymes of purine catabolism; uricase and urease

Uricase and urease, the first and last enzymes in the uricolytic pathway, are both present in the tissues of *G. natalis* (Table 1). It is likely, therefore, that the intermediate enzymes of the uricolytic pathway, allantoinase and allantoicase, are also present. The uricolytic pathway appears to be split between tissues, since activities of the first enzyme, uricase, were highest in the spongy connective tissue (branchiostegite) while activities of the last enzyme, urease, were highest in the gills (Table 1). This pattern of enzyme activity suggests that urate catabolism begins in the spongy connective tissue cells where the urate is stored (Linton and Greenaway 1997a). Within these cells, the urate could be degraded to either allantoin, allantoic acid or urea. Urea is the most likely candidate since it is present, sometimes at high concentrations (0.5–10.77 mmol l^{-1}), in the blood of terrestrial crabs (Greenaway and Nakamura 1991; DeVries *et al.* 1994; Taylor and Greenaway 1994; Linton and Greenaway 1995). After initial catabolism of urate, the product (most likely urea) is probably transported, via the haemolymph, to the gills where urease catabolises it to ammonia. The occurrence of uricase activity in the gills of a small proportion of the animals assayed [37.5% of animals assayed (Table 1)] indicates that complete urate catabolism could possibly occur in the gills of some animals. As the gills are the site of ammonia excretion in *G. natalis* (Greenaway and Nakamura 1991), the ammonia produced by the hydrolysis of urea is probably excreted. Given the presence and location of the uricolytic enzymes in the tissues of *G. natalis* it is clear firstly that the crab is potentially able to catabolise its urate reserves and secondly that the urate deposits could be a temporary store of nitrogenous wastes (Table 1). In *G. natalis* the activity of uricase is less than that of urease and like uricase in the crayfish *O. rusticus* and the terrestrial isopod *O. asellus* it may also be the rate limiting enzyme of uricolysis (Fig. 1; Table 1) (Hartenstein 1968; Sharma and Neveu 1971).

Possible physiological functions of the urate deposits

Terrestrial Crustacea in positive water balance have efficient mechanisms for the excretion of nitrogenous wastes and it is unlikely that they need to store them. Gecarcinid crabs excrete ammonia via the excreted fluid while terrestrial isopods such as *Porcellio scaber*, the grapsid *Geograpsus grayi* and to some extent the ghost crab *Ocypode quadrata* excrete ammonia gas (reviewed Greenaway 1991; O'Donnell and Wright 1995). In certain situations, however, excretion of nitrogenous wastes as ammonia is not possible. During dehydration, urine flow in gecarcinid crabs is curtailed and the usual route for excretion of ammonia is blocked (Wood *et al.* 1986; Greenaway 1991). Similarly, terrestrial isopods such as *P. scaber* in humidities that are below the threshold for water vapour absorption, do not produce pleon fluid for water vapour absorption and are unable to excrete their nitrogenous wastes as ammonia gas (Wright and O'Donnell 1992, 1993; O'Donnell and Wright 1995). Also amphibious crabs such as *C. hirtipes*, *Holthuisana tranversa* and *Potamonautes warreni* without water to immerse themselves are unable to excrete ammonia from the gills into the surrounding water (Linton and Greenaway 1995; Dela-Cruz and Morris 1997; Morris and Van Aardt 1998). The time period required for such storage would vary. Isopods may require storage for approximately 12–14 h while they are foraging, nocturnally, in humidities below the threshold for water vapour absorption (Wieser *et al.* 1969; Kirby and Harbaugh 1974; O'Donnell and Wright 1995). In contrast, *H. transversa* living in drought conditions may not have access to free water for periods in excess of 1 year and would need to store their nitrogenous wastes during this time (Linton and Greenaway 1995). Amphibious gecarcinid crabs such as *C. hirtipes* and *C. carnifex* during experimental water deprivation have been reported to store their nitrogenous wastes for up to 9 days (Wood *et al.* 1986; Dela-Cruz and Morris 1997).

In the situations described above, the animals must store their nitrogenous wastes to prevent toxicity. Urate is a likely candidate for such storage since as a rounded solid it is non-toxic, does not contribute to the osmolarity of the cell, stores nitrogen in a concentrated anhydrous form and as mentioned is relatively cheap to synthesise. Amino acids such as glutamate, glutamine and glycine have also been suggested as temporary detoxification and storage molecules for nitrogenous wastes in terrestrial Crustacea (reviewed Greenaway 1991; O'Donnell and Wright 1995). Use of amino acids for such storage could be only viable for short periods (hours) since amination of large amounts of α-ketoglutarate would drain intermediates from the Krebs cycle and hence impede ATP production (Stryer 1988). The presence of large amounts of amino acids in the cells could also hinder osmoregulation and ion transport as amino acids in solution contribute to the osmolarity and if charged (e.g. glutamate) alter electrical gradients.

The stored urate could also be degraded and the nitrogen released incorporated into amino acids and hence reutilised. The midgut gland is the most likely tissue in which this might occur since it is involved in assimilation and digestion of protein and hence amino acid metabolism (Icely and Nott 1992). Given that uricase and urease activities were rarely detected in the midgut gland of G. *natalis* (Table 1) and animals on either a nitrogen-free diet or during situations of high nitrogen requirement such as oogenesis did not degrade urate to compensate for dietary nitrogen deficiencies (S. Linton unpublished data) it is unlikely that the midgut gland performs this role.

Acknowledgements. This work was supported by an Australian Research Council grant 942611 (P.G.) and by an Australian Postgraduate award (S.L.).

References

Battelli, M.G., Della Corte, E., Stirpe, F. (1972) Xanthine oxidase type D (Dehydrogenase) in the intestine and other organs of the rat. *Biochem. J.* **126**:747–749

Campbell, J.W. (1995) Excretory nitrogen metabolism in reptiles and birds. In Walsh, P.J., Wright, P. (eds) *Nitrogen metabolism and excretion*. CRC, Boca Raton, Fla., pp. 147–178

Chappell, D.J., Slaytor, M. (1993) Uric acid synthesis in freshly collected and laboratory-maintained *Nasutitermes walkeri* Hill. Inset. *Biochem. Mol. Biol.* **23**:499–506

Claybrook, D.L. (1983) Nitrogen metabolism. In Mantel, L.H. (ed) *The biology of Crustacea*, vol 5. Academic, New York, pp. 163–213

Dela-Cruz, J., Morris, S. (1997) Water and ion balance and nitrogenous excretion as limitations to terrestrial excursion in the Christmas Island Blue Crab, *Cardisoma hirtipes. J. Exp. Zool.* **279**:537–548

Della Corte, E., Stirpe, F. (1968) Regulation of xanthine oxidase in rat liver: modifications of the enzyme activity of rat liver supernatant on storage at -20°C. *Biochem. J.* **108**:349–351

Della Corte, E., Stirpe, F. (1972) The regulation of rat live xanthine oxidase. Involvement of thiol groups in the conversion of the enzyme activity from dehydrogenase (Type D) into oxidase (Type O) and purification of the enzyme. *Biochem. J.* **126**:739–745

De Vries, M.C., Wolcott, D.L., Holliday, C.W. (1994) High ammonia and low pH in the urine of the ghost crab, *Ocypode quadrata. Biol. Bull.* **186**:342–348

Dykens, J.A. (1991) Purineolytic capacity and origin of haemolymph urate in *Carcinus maenas* during hypoxia. *Comp. Biochem. Physiol.* **98B**:579–582

Farrelly, C.A., Greenaway, P. (1993) Land crabs with smooth lungs: Grapsidae, Gecarcinidae, and Sundathelphusidae ultrastructure and vasculature. *J. Morphol.* **215**:245–260

Fischer, E. (1926) L'Hépato-pancréas des Crustacés excrète des substances puriques. *CR Séances Soc. Biol. Fil.* **95**:1476–1478

Gifford, C.A. (1968) Accumulation of uric acid in the land crab, *Cardisoma guanhumi*. *Am. Zool.* **8**:521–528

Graf, F., Michaut, P. (1975) Chronologie du développement et évolution du stockage de calcium et des cellules à urates chez *Niphargus schellenbergi* Karaman. *Int. J. Speleol.* **7**:247–272

Greenaway, P. (1991) Nitrogenous excretion in aquatic and terrestrial crustaceans. *Mem. Queensl. Mus.* **31**:215–227

Greenaway, P., Morris, S. (1989) Adaptations to a terrestrial existence by the robber crab, *Birgus latro* L. III Nitrogenous excretion. *J. Exp. Biol.* **143**:333–346

Greenaway, P., Nakamura, T. (1991) Nitrogenous excretion in two terrestrial crabs *Gecarcoidea natalis* and *Geograpsus grayi*. *Physiol. Zool.* **64**:767

Harrison, J.F. (1995) Nitrogen metabolism and excretion in locusts. In Walsh, P.J., Wright, P. (eds) *Nitrogen metabolism and excretion*. CRC, Boca Raton, Fla., pp. 119–147

Hartenstein, R. (1968) Nitrogen metabolism in the terrestrial isopod, *Oniscus asellus*. *Am. Zool.* **8**:507–519

Horne, F.R. (1968) Nitrogen excretion in Crustacea. I. The herbivorous land crab *Cardisoma guanhumi* (Latreille). *Comp. Biochem. Physiol.* **26**:687

Icely, J.D., Nott, J.A. (1992) Digestion and absorption: digestive system and associated organs. In: Harrison, F.W., Humes, A.G. (eds) *Microscopic anatomy of invertebrates*, vol 10, *Decapod Crustacea*. Wiley-Liss, New York, pp. 529–617

Johnson, P.T. (1980) This histology of the blue crab *Callinectes sapidus*, a model for the decapoda. Praeger, New York, pp. 1–440

Kirby, P.K., Harbaugh, R.D. (1974) Diurnal patterns of ammonia release in marine and terrestrial isopods. *Comp. Biochem. Physiol.* **47A**:1313–1321

Kooij, A. (1994) A re-evaluation of the tissue distribution and physiology of xanthine oxidoreductase. *Histochem. J.* **26**:889–915

Lallier, F.H., Walsh, P.J. (1991) Activities of uricase, xanthine oxidase, and xanthine dehydrogenase in the hepatopancreas of aquatic and terrestrial crabs. *J. Crust. Biol.* **11**:506–512

Linton, S.M., Greenaway, P. (1995) Nitrogenous excretion in the amphibious crab *Holthuisana transversa* under wet and dry conditions. *J. Crust. Biol.* **15**:633–644

Linton, S.M., Greenaway, P. (1997a) Intracellular purine deposits in the gecarcinid land crab *Gecarcoidea natalis*. *J. Morphol.* **231**:101–110

Linton, S.M., Greenaway, P. (1997b) Urate deposits in the gecarcinid land crab *Gecarcoidea natalis* are synthesised *de novo* from excess dietary nitrogen. *J. Exp. Biol.* **200**:2347–2354

Liras, A., Rotllan, P., Llorente, P. (1992) *De novo* purine biosynthesis in the crustacean Artemia: influence of salinity and geographical origin. *J. Comp. Physiol.* **162B**:263–266

McCord, J.M. (1985) Oxygen-derived free radicals in postischemic tissue injury. *New Engl. J. Med.* **312**:159–163
Morris, S., Van Aardt, W.J. (1998) Salt and water relations, and nitrogen excretion, in the amphibious African freshwater crab *Potamonautes warreni* in water and in air. *J. Exp. Biol.* (in press)
Noguchi, T., Takada, Y., Fujiwara, S. (1979) Degradation of uric acid to urea and glyoxylate in peroxisomes. *J. Biol. Chem.* **254**:5272–5275
O'Donnell, M.J., Wright, J.C. (1995) Nitrogen excretion in terrestrial crustaceans. In: Walsh, P.J., Wright, P. (eds) *Nitrogen metabolism and excretion.* CRC, Boca Raton, Fla., pp. 105–118
Sharma, M.L., Neveu, M.C. (1971) Studies of the role of ornithine cycle and purine catabolism in urea biosynthesis in the fresh water crayfish, *Orconectes rusticus. Comp. Biochem. Physiol.* **40B**:863–870
Stryer, L. (1988) *Biochemistry*, 3rd edn. Freeman, New York
Taylor, H.H., Greenaway, P. (1994) The partitioning of water and solutes during evaporative water loss in three terrestrial crabs. *Physiol. Zool.* **67**:539
Wägele, J.W. (1992) Isopoda. In: Harrison, F.W., Humes, A.G. (eds) *Microscopic anatomy of invertebrates*, vol 9, *Crustacea.* Wiley-Liss, New York, pp. 529–617
Wieser, W., Schweizer, G., Hartenstein, R. (1969) Patterns in the release of gaseous ammonia by terrestrial isopods. *Oecologia* **3**:390–400
Wolcott, D.L., Wolcott, T.G. (1984) Food quality and cannibalism in the red land crab *Gecarcinus lateralis. Physiol. Zool.* **57**:318–324
Wolcott, D.L., Wolcott, T.G. (1987) Nitrogen limitation in the herbivorous land crab *Cardisoma guanhumi. Physiol. Zool.* **60**:262–268
Wood, C.M., Boutilier, R.G., Randall, D.J. (1986) Osmoregulation, ionic exchange, blood chemistry and nitrogenous waste in the land crab *Cardisoma carnifex:* a field and laboratory study. *Biol. Bull.* **169**:267–290
Wright, J.C., O'Donnell, M.J. (1992) Osmolality and electrolyte composition of pleon fluid in *Porcellio scaber* (Crustacea, Isopoda, Oniscidea): implications for water absorption. *J. Exp. Biol.* **164**:189–293
Wright, J.C., O'Donnell, M.J. (1993) Total ammonia concentration and pH of haemolymph, pleon fluid and maxillary urine in *Porcellio scaber* Lattreille (Isopoda, Oniscidea): relationships to ambient humidity and water vapour uptake. *J. Exp. Biol.* **176**:233–246

Flight metabolism in carpenter bees and primary structure of their hypertrehalosaemic peptide

Gerd Gäde · Lutz Auerswald

Zoology Department, University of Cape Town, Rondebosch 7701, South Africa

Received: 27 March 1998 / Accepted: 13 May 1998 / Published: 18 June 1998

Abstract. We measured the rate of oxygen consumption and carbon dioxide production as well as energy substrates in haemolymph and flight muscles of carpenter bees of the genus *Xylocopa* at rest and after tethered lift-generating flight. Flight of 2 min duration at an ambient temperature of 28°C elevated oxygen consumption about 70-fold above resting rate. The respiratory quotient during rest and flight was 1 indicating that carbohydrates were the exclusive substrate oxidised. This was corroborated by measurements of metabolism. Carbohydrates are in high concentrations in the haemolymph. This store was significantly diminished during a 10-min flight period. Whereas lipids did not contribute to energy provisions, the proline concentration in the haemolymph and in the flight muscles was significantly decreased upon flight, but the amount can only account for a very small contribution to overall flight metabolism. Polysaccharide reserves in flight muscles and whole abdomina are almost non-existent. However, earlier studies had identified the crop as a source of oligosaccharides (Louw and Nicolson 1983). Carbohydrate metabolism is influenced by a metabolic peptide from the corpus cardiacum. We could isolate a peptide from the corpora cardiaca of carpenter bees, which by retention time in HPLC and by its mass is very likely characterised as the octapeptide Scg-AKH-II (pGlu-Leu-Asn-Phe-Ser-Thr-Gly-Trp-NH_2) previously shown to occur in certain Orthoptera. This is the first member of the large AKH/RPCH family of peptides to be identified from a hymenopteran species. Injection of the synthetic peptide into adult carpenter bees caused carbohydrate mobilisation. We suggest that the peptide targets the high sugar stores in the crop and speculate that it may facilitate sugar passage rate through the digestive system.

Key words. AKH/RPCH · Carpenter bee · Energy metabolism · Insect flight Neuropeptide · Xylocopa

Correspondence to: G. Gäde (e-mail: ggade@botzoo.uct.za, Fax: +27-21-6503301)

Introduction

The ability to fly is one of the most conspicuous characteristics of a large number of insects. Flight requires huge amounts of energy and, hence, it is not surprising that active flight muscles of insects show the highest rate of energy turnover known in nature. Since these muscles perform their work completely aerobically, extreme respiratory rates have been measured; for example, rates of oxygen consumption of up to 150 ml O_2 g^{-1} body mass h^{-1} during free hovering flight (Casey 1989).

Fuels utilised during flight can differ quite drastically in the various groups of insects. Characteristically, Lepidoptera and Orthoptera burn carbohydrates together with lipids; however, especially in moth species that do not feed during the adult stage and during sustained flight in locusts, only lipids are utilised (see Sacktor 1975; Beenakkers et al. 1981; Goldsworthy 1983; Wheeler 1989; Gäde 1992). In tsetse flies and certain dung beetle species proline oxidation appears to be the sole fuel for flight, whereas other beetles of the families Scarabaeoideae, Chrysomelidae and Meloidae oxidise proline in conjunction with carbohydrates (Bursell 1981; Gäde and Auerswald 1998). Most species of the orders Diptera and Hymenoptera exclusively oxidise carbohydrates during flight (Sacktor 1975; Beenakkers et al. 1981; Wheeler 1989; Gäde 1992). Although we have a good knowledge of flight energetics in honey bees and bumble bees, the physiology and energetics of flight in carpenter bees is less well known (Nicolson and Louw 1982). Although there is no or little storage of glycogen in honey bees and bumble bees, the digestive tract, including the crop, contains the necessary chemical energy in the form of non-polymeric carbohydrates (Surholt et al. 1988; Panzenböck and Crailsheim 1997). Both groups of Hymenoptera exclusively use carbohydrates for energy provision during flight as evidenced by respiratory gas measurements resulting in a respiratory quotient of 1 (Bertsch 1984; Rothe and Nachtigall 1989).

In Sub-Sahara Africa there are no bumble bees. Instead, carpenter bees (Xylocopidae) are found. Two fairly large species of carpenter bees, *Xylocopa capitata* and *Xylocopa caffra*, which fly very well, are quite common in the Western Cape province of South Africa. They feed on nectar and pollen produced by the flowers of Papilionaceae and Labiatae. *X. capitata* depends almost exclusively in this region on the blossoms of the *Virgilia* tree (Fabaceae) (Louw and Nicolson 1983). It is therefore anticipated that carpenter bees have a metabolism comparable to that of honey bees and bumble bees. However, until no experimental data are present, it cannot be argued with certainty. The present study, therefore, analyses the fuels during flight and uses a method of tethered flight in which the insect produces lift and which is assumed to closely resemble the situation during free flight in nature (see Auerswald *et al.* 1998).

Moreover, we were also interested to characterise putative substrate-mobilising neuropeptides of the adipokinetic/red pigment-concentrating hormone (AKH/RPCH) family present in the corpora cardiaca from carpenter bees, since no member of the family in Hymenoptera has been identified yet (see Gäde 1997a; Gäde et al. 1997).

Materials and methods

Insects

Most experiments were performed on adult carpenter bees of both sexes of the species *X. capitata* Smith with an average body mass of 1.3±0.2 g. In addition, for some experiments specimens of *X. caffra Linnaeus* with an average body mass of 0.75±0.2 g were used. For comparison, some males of the latter species were also used to measure metabolites in body parts. Both species were caught near the campus of the University of Cape Town and at Kirstenbosch Botanical Gardens by netting when they were feeding on flowers of various Papilionaceae or Labiatae. The physiological state and age of the bees were not known.

Insects were kept in transparent 2 l containers with flowers of their food plant and with access to water at 24–26°C for 16 h. Alternatively, corpora cardiaca were dissected immediately after return to the laboratory.

Flight

Only carpenter bees which were absolutely quiescent in the container, i.e. did not walk around or tried to fly, were used. Experiments were performed at 24–25°C in a constant temperature room. The carpenter bee was attached via the thorax to a bent insect pin using dental wax. The pin (mass 85 mg) could move inside a piece of metal tube (diameter 2 mm) in such a way that it was possible to detect whether the insect produced lift as happens during natural flight. This is shown in Clip 1 where a female *X. caffra* is used in flight experiments. When the top of the insect pin moves up, the insect is considered to be producing lift. In Clip 1 the pin is pushed back with a finger to demonstrate that the carpenter bee will produce lift immediately again and the pin does rise again. The apparatus was held in a wooden clamp mounted on a stand. In addition, this method allows the insect to turn on its vertical and horizontal axes. For more details, see Auerswald et al. (1998).

After the carpenter bee had been attached to the pin, a 1 μl haemolymph sample was taken from the intersegmental membrane between the thorax and the abdomen for the determination of metabolite concentrations. After attachment to the stand, the carpenter bee was provided with a Styrofoam ball for tarsal contact. The insect remained quiescent until it displayed pumping movements with the abdomen and started to fly thereafter. A second 1 μl haemolymph sample was taken immediately after the end of a 10 min flight, and a third sample was taken after a 2 h period of rest following the 10 min of flight. The haemolymph samples were either blown into 100 µl of acetonitrile (80%) for amino acid determination or into 100 µl sulphuric acid (conc.) for determination of lipids and carbohydrates (see below).

Whole thoraces or abdomina were separated from the rest of the body, wrapped in aluminium foil, and frozen in liquid nitrogen. The procedure took no more than 1 min. Samples were kept frozen at -20°C until further use.

Bioassay

Two hours prior to experimentation, carpenter bees were transferred into small black containers (50 ml). For bioassays, they were taken out of the container and wrapped in tin foil in such a way that the intersegmental membrane was exposed and the thorax and the abdomen could be bent to allow haemolymph sampling. After 1 µl of haemolymph had been taken and blown into 100 µl of sulphuric acid for determination of carbohydrate concentration, either distilled water or the synthetic peptide Scg-AKH-II was injected laterally into the abdomen. The insect was returned to the container and 2 h later a second 1 µl haemolymph sample was taken.

Analyses of metabolites

As described previously (Auerswald et al. 1998), 1 µl of haemolymph was taken and analysed either for the concentration of total lipids and carbohydrates or for the levels of the amino acids proline and alanine. Frozen thoraces and abdomina were broken apart under liquid nitrogen and gut and chitin removed. The remaining tissue was ground to a powder in liquid nitrogen, extracted with perchloric acid, and polysaccharides as well as the amino acids proline and alanine were determined (Zebe and Gäde 1993).

Determination of the rate of oxygen consumption and respiratory quotient

An open flow-through respiratory system consisting of an S-3A oxygen analyser (Applied Electrochemistry, Sunnyvale, Calif., USA) and a 225 MK3 infrared gas analyser (ADC, UK) to measure changes in the oxygen and carbon dioxide concentrations, respectively, was used at 28±1°C. In the respiration chamber, the carpenter bees were attached to an insect pin using the method described above (for more details, see Auerswald et al. 1998).

Isolation and identification of peptides from corpora cardiaca

Glands were dissected into 80% methanol and extracts were prepared as described previously (Gäde *et al.* 1984). The extract was applied to a Nucleosil C-18 column for reversed-phase high performance liquid chromatography as

Clip 1. For experiments investigating metabolite changes during lift-generating tethered flight, carpenter bees were attached to an insect pin using dental wax. The insect was considered to produce lift when the top of the pin went up. Here a female *X. caffra* is shown

Clip 2. Female carpenter bee (*X. caffra*) collecting nectar from flowers

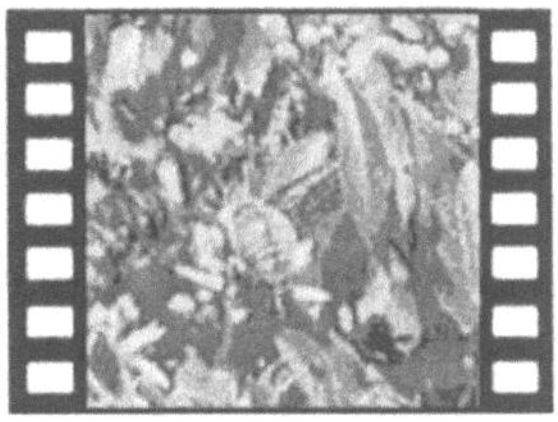

Clip 3. Male carpenter bee (*X. caffra*) feeding on flowers

Clip 4. Male carpenter bee displaying territorial behaviour. The bee is hovering in the centre of its territory and leaves frequently for short patrolling flights

outlined previously (Gäde 1985; see also Fig. 3). The purified fraction, which was active in the bioassay, was used to generate spectra of mass data using a Voyager Elite (PerSeptive Biosystems, Framingham, Mass., USA) matrix-assisted laser desorption/ionisation instrument. Samples were prepared in alpha-cyano-4-hydroxycinnamic acid, and spectra acquired in the positive mode.

Synthetic peptides

Adipokinetic hormones of the migratory (*Locusta migratoria*) and desert (*Schistocerca gregaria*) locust, code-named Lom-AKH-I, Lom-AKH-II and Scg-AKH-II, adipokinetic hormone from the tobacco hornworm moth (*Manduca sexta*; Mas-AKH) and the hypertrehalosaemic peptides from the American cockroach (*Periplaneta americana*; Pea-CAH-I and Pea-CAH-II) came from Peninsula Laboratories (Belmont, Calif., USA).

Results

Foraging and territorial behaviour

Female *X. caffra*, which are recognisable by their dark colour with yellow stripes, were busy foraging throughout the day and stopped flying only when they were collecting nectar from a flower. They did not display any long hovering periods. A short sequence of a typical foraging flight by the female *X. caffra* is shown in Clip 2. In contrast, males of the same species, which are recognisable by their smaller size as well as by their yellow fur, were only feeding in the morning hours until midday in a similar manner as the females (Clip 3). Later in the day, however, the males displayed an increasingly territorial behaviour and were hovering in the centre of their territory (Clip 4). This was usually under a branch of a tree or bush. Periodically, the male carpenter bee left this place for short flights to patrol its territory and returned thereafter to the same place under a branch from where they had started their journey (Clip 4). When a male competitor of the same species entered the territory, the two males fought for several minutes during which they performed hovering flight facing one other. In contrast, the males of *X. capitata* interrupt their flight frequently by short periods of wing-buzzing (Louw and Nicolson 1983).

Metabolite changes during flight

The level of carbohydrates in the haemolymph of resting *X. capitata* was high at 36.8±11.4 mg ml^{-1} and decreased significantly to 23.5±12.0 mg ml^{-1} during 10 min of flight (Fig. 1). After 2 h of rest following a 10 min flight, the carbohydrate concentration in the haemolymph had not changed and was at the same level as after flight.

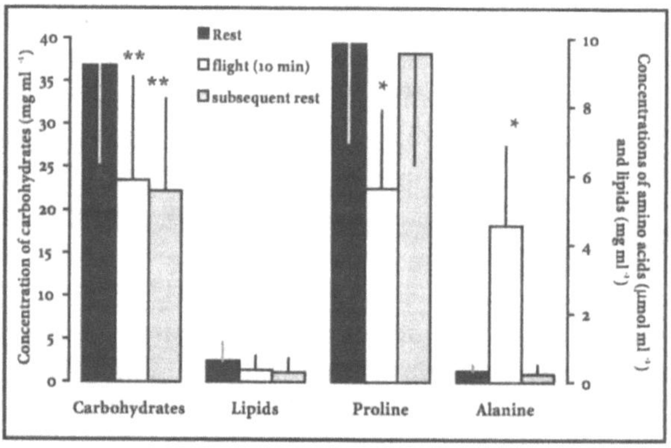

Fig. 1. Changes of some metabolites in the haemolymph of *X. capitata* during flight. The concentrations of lipids (n=14), carbohydrates (n=14) and of the amino acids proline (n=6) and alanine (n=6) are given as means ± S.D. during rest, after 10 min of lift-generating flight and 2 h of rest after 10 min of flight. Significance levels of differences compared with resting concentration: *P<0.02, **P<0.005 using paired t-test

The concentration of proline in the haemolymph of resting *X. capitata* was low at 9.9±2.9 μmol ml^{-1} but decreased significantly to 5.6±2.3 μmol ml^{-1} during 10 min of lift-generating tethered flight (Fig. 1). Concomitantly, the level of alanine increased significantly from 0.3±0.2 to 4.6±2.3 μmol ml^{-1}. During a resting period of 2 h following 10 min of flight, the initial concentrations of both amino acids were re-established.

The concentration of lipids in the haemolymph of resting *X. capitata* was low and did not change significantly during flight and subsequent rest (Fig. 1).

The concentration of glycogen (more precisely polysaccharides) in the flight muscles of *X. caffra* at rest was barely measurable at 0.54±0.35 μmol glucose equivalent g^{-1} fresh mass and did not change during flight nor flight and subsequent rest (Fig. 2).

The concentration of proline in the flight muscles at rest was low at 3.7±0.2 μmol g^{-1} fresh mass; there was a small but significant decrease to 3.1±0.6 μmol g^{-1} during a flight period of 10 min. The alanine level increased significantly from 0.3±0.1 μmol g^{-1} at rest to 1.2±0.7 μmol g^{-1} after flight (Fig. 2). Initial concentrations of proline and alanine were restored during the rest period after flight.

Because the levels of glycogen were very low in the flight muscles, we searched for polysaccharide reserves in abdomina of female and male carpenter bees. Resting females of *X. caffra* had an abdominal concentration of 2.0±0.9 μmol glucose equivalents g^{-1} fresh mass (n=5), whereas the males had hardly any reserves in their abdomina (0.1±0.1 μmol glucose equivalents g^{-1}; n=4). The resting proline concentration in abdominal tissue of female (n=5) and male (n=4) *X. caffra* was 2.6±1.6 and 2.7±1.1 μmol g^{-1}, respectively; the alanine concentration was 1.8±1.4 μmol g^{-1} for females and 0.9±1.1 μmol g^{-1} for males.

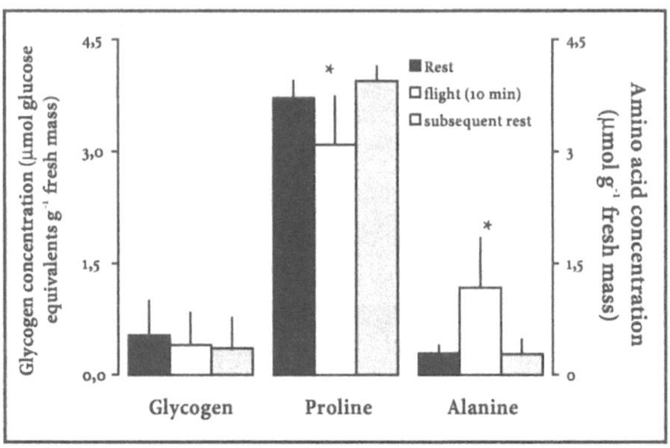

Fig. 2. Changes of the concentrations of glycogen, proline and alanine in flight muscle tissue of *X. caffra* at rest, after 10 min of tethered lift-generating flight and recovery following 10 min of flight. Values are means ± SD (n=5–6). Significance levels of differences compared with resting concentration: *P<0.05 using Student's t-test

Oxygen consumption and respiratory quotient during flight

The oxygen consumption of resting *X. capitata* at 28°C was 1.1±0.2 ml g^{-1} h^{-1} (n=4) and the respiratory quotient was calculated to be 1.01±0.1 (n=4). Lift-generating tethered flight for 2 min increased the oxygen consumption about 70-fold to 75.0±5.3 ml g^{-1} h^{-1} (n=4). The respiratory quotient value was 1.04±0.07 (n=4).

Peptide purification and characterisation

The UV elution profiles of a methanolic extract of corpora cardiaca from the two carpenter bee species *X. capitata* (ten gland equivalents, Fig. 3A) and *X. caffra* (seven gland equivalents, Fig. 3B) on an analytical RP-C-18 column both show a distinct UV absorbance peak at 12.4 min, which corresponds to a fluorescence peak when monitoring for the characteristics of tryptophan, a conserved residue at position 8 in the sequence of peptides of the AKH/RPCH family (Fig. 3C,D). Because the retention time of the distinct peak was close to those known for the peptides of certain other insects, we injected the synthetic peptides Pea-CAH-I and II, Lom-AKH-I and II, Scg-AKH-II and Mas-AKH as well onto the analytical column and monitored their retention times. Scg-AKH-II was eluted at an identical retention time (12.4 min) as the peak from the glands of the carpenter bee species (see Fig. 3D). When co-injected, car-

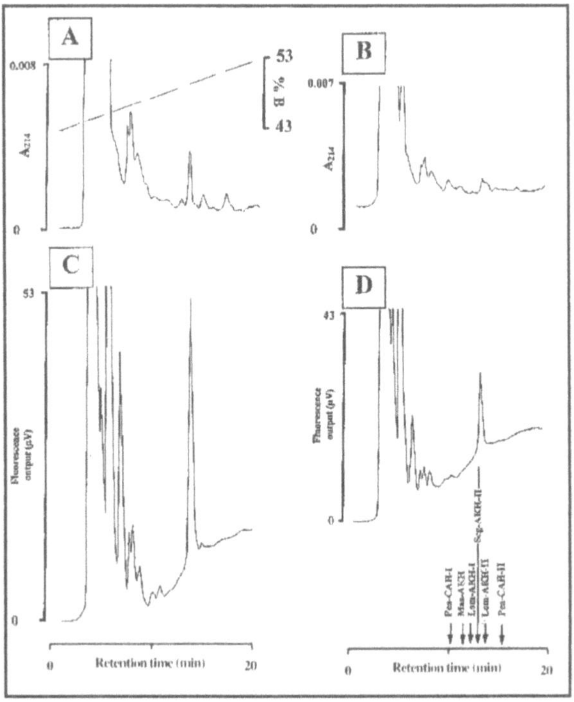

Fig. 3. Extracts of ten corpora cardiaca from *X. capitata* and seven corpora cardiaca of *X. caffra* were run on HPLC and the UV absorbance (**A**: *X. capitata*, **B**: *X. caffra*) and the fluorescence signal (**C**: *X. capitata*, **D**: *X. caffra*) recorded. *Arrows* indicate retention times of various synthetic peptides of a control run

penter bee peak material and Scg-AKH-II were co-eluted (data not shown). The material from the carpenter bee fraction had hypertrehalosaemic activity when injected into the bees (data not shown). When peak material and synthetic Scg-AKH-II were both analysed by MALDI mass spectrometry, each had mass peaks at *m/z* 956 and 972 which correspond with the theoretical value for the sodium and potassium cationised molecule ($[M+Na]^+=956$; $[M+K]^+=972$). This makes it very likely that the endogenous peptide in *Xylocopa* corpora cardiaca has the primary structure pGlu-Leu-Asn-Phe-Ser-Thr-Gly-Trp-NH_2, i.e. the same structure as Scg-AKH-II.

Effect of Scg-AKH-II on carbohydrate levels in haemolymph

Upon injection of 20 pmol Scg-AKH-II, the carbohydrate concentration in the haemolymph of *X. capitata* increased significantly from the initial level of 46.0±12.8 to 61.0±22.0 mg ml^{-1} while no significant change was observed after injection of distilled water (Table 1).

Discussion

The present study was undertaken to unravel the question of the main fuel source during flight of carpenter bees and, additionally, to characterise a putative neuropeptide responsible for metabolic fuel mobilisation.

Flight metabolism of carpenter bees

The data provided show unequivocally that carbohydrates are the major fuel during flight. There are three lines of evidence. Firstly, carbohydrates are quantitatively the main source of oxidisable energy stored in the haemolymph of resting carpenter bees. The concentrations are high with about 40 mg/ml, which is in the same range found in the honey bee (Woodring et al. 1994), but a 50% lower level was reported for another carbohydrate flyer, the blowfly *Phormia terraenovae* (Wilps and Gäde 1990). The lipid levels in the haemolymph of carpenter bees are very low, and this is true for the proline concen-

Table 1. Hypertrehalosaemic response in *X. capitata* upon injection of synthetic Scg-AKH-II. Values are given as mg ml^{-1} (mean ± SD)

	n	Before	After 120 min	Difference
Distilled water	8	38.0±13.3	44.5±22.4	6.5±19.5
Scg-AKH-II (20 pmol)	10	46.0±12.8	61.0±22.0	15.0±13.0*

*Significantly different ($P<0.02$) using paired *t*-test

tration as well. Beetles, for example, which use proline in conjunction with carbohydrates or exclusively as fuel for flight, contain between 35 µmol ml^{-1} (*Decapotoma lunata*; Auerswald and Gäde 1995), 70 µmol ml^{-1} (*Scarabaeus rugosus*; Gäde 1997b) and 130 µmol ml^{-1} (*Pachnoda sinuata*; Auerswald and Gäde 1995), whereas we measured only 10 µmol ml^{-1} in the carpenter bee.

Secondly, upon lift-generating flight for 10 min the only metabolites that are used considerably in the carpenter bee are the haemolymph carbohydrates. Lipids do not contribute and the changes in the concentration of proline, although significant, are not very important overall for flying (but see below).

Thirdly, respiratory gas measurements clearly show that the respiratory quotient is 1 at rest and during flight. Thus, when carpenter bees are inactive or when their metabolism is increased by a factor of 70, as evidenced by the increased rate of oxygen consumption during flight, carbohydrates are oxidised. This has been shown also for their „close relatives", the bumble bees (Bertsch 1984) and honey bees (Rothe and Nachtigall 1989). Our measurement for the rate of oxygen consumption during flight, 75 ml O_2 g^{-1} h^{-1}, is higher than previously reported (52 and 66 ml O_2 g^{-1} h^{-1} for *X. capitata* and *X. californica*, respectively; Chappell 1982; Nicolson and Louw 1982), suggesting that our method of tethered but lift-generating flight is measuring maximal power output. A similar mass-specific rate of oxygen consumption as for carpenter bees, namely between 80 and 95 ml O_2 g^{-1} h^{-1}, was found for bumble bees (Heinrich 1975) and honey bees (Withers 1981) during flight.

In accordance with previous data on bumble bees and honey bees (Surholt et al. 1988; Panzenböck and Crailsheim 1997), carbohydrates are not stored in the form of polysaccharides in flight muscles or abdomen of carpenter bees. This is quite intriguing. As can be seen in the video clips, carpenter bees are powerful flyers and do so for long periods of time with only very short periods of feeding stops. Especially the males of *X. caffra* are engaged in hovering flight for a significant period of time during their territorial behaviour. Such a flight mode is most costly. It is clear that carbohydrates from the haemolymph are not sufficient to support the energy for muscle contraction during such activity. That leaves two possibilities: either there are other reserves stored or the carpenter bees forage sufficient amounts of nectar. Both areas have been analysed previously (Louw and Nicolson 1983). According to their study on *X. Capitata*, foraging females and territorial males carry a highly concentrated sugar solution (750 mg ml^{-1}) in their crop. Moreover, these authors calculated an energy budget for foraging females where nectar from two out of the eight visited flowers per minute will be used to „repay" the energy costs for flying, while the difference is stored in the crop.

Interestingly, we did not find a replenishment of the carbohydrate concentration in the haemolymph 2 h after a flight of 10 min. This indicates that during the flight period, when carpenter bees could not feed in our experimental set-up, the crop reserves were used up and were not available after the flight for restoring the carbohydrate level in the haemolymph. In the next austral spring experiments could be done checking crop reserves during natural flight with intermitting stops for feeding versus laboratory flight without feeding.

The levels of proline warrant one comment. Although there is no clear indication that the amino acid is used to power the flight muscles, the significant and roughly equimolar changes of proline and its partial oxidation product alanine during flight rather seem to suggest that it is used to „spark" the Krebs cycle as proposed for the blowfly *Phormia regina* (Sacktor and Childress 1967). However, it was recently found that proline also fulfills a role during endothermic warm-up of the African fruit beetle *Pachnoda sinuata* (Auerswald et al. 1998). During our flight experiments with carpenter bees we observed pumping movements of the abdomen indicative of endothermic heat production prior to flight activity. Carpenter bees need a flight muscle temperature of at least 37°C (Nicolson and Louw 1982). Thus, another area of future research may be to establish whether in the preparatory phase before take-off proline may serve for energy provision.

The putative energy-mobilising neuropeptide

Hypertrehalosaemic factors have been shown to occur in crude extracts of corpora cardiaca from the honey bee and a wasp (Woodring et al. 1994). These factors were effective in *Periplaneta americana* and in the Hymenoptera itself. Although isolation was successfully attempted, no structural information is available to date. Here we report that the hypertrehalosaemic factor in the corpora cardiaca from carpenter bees of the genus *Xylocopa* can very likely be attributed to a peptide of the large AKH/RPCH family. Although we have not sequenced the peptide, the current evidence is strong. Retention time and mass data are identical to those for the known peptide Scg-AKH. No D-amino acids are known to occur in AKH/RPCH family peptides and if, for example, an isoleucine residue would reside in the molecule instead of leucine (both have same mass of 113), then a different retention time would result. This is clearly shown for the peptides code-named Pea-CAH-II and Poa-HrTH from cockroaches which contain Leu and Ile at position 2 respectively (Gäde 1989). Thus, we feel strongly that carpenter bees have the peptide Scg-AKH-II which was first found in the locusts *Schistocerca gregaria* and *S. nitans* (Siegert et al. 1985; Gäde et al. 1986), but has since been found in pyrgomorphid grasshoppers and certain Ensifera (Gäde 1996; Gäde et al. 1997). It is the first time, thus, that this neuropeptide has been characterised from a non-orthopteran insect. Injection of synthetic Scg-AKH-II into carpenter bees elevates the carbohydrate level in the haemolymph. The question arises as to where the extra carbohydrate is mobilised from when no stores are available in the abdomen. As observed by Louw and Nicolson (1983) in a study of carpenter bees captured during late austral spring, a visual inspection did not reveal the presence of any fat body tissue in carpenter bees of the present study. This is, apparently, different to newly emerged carpenter bees which have well-developed fat bodies (Louw and Nicolson 1983). The extra carbohydrate resulting in the observed higher concentration after injection of Scg-AKH-II can be explained by mobilisation of stored sugars from the crop. This, then, points to a possible new action of the neuropeptide: it is not responsible for the activation of glycogen phosphorylase as discussed for other hypertrehalosaemic peptides (Gäde 1996), but may facilitate the passage of oligosaccharides from crop via digestive tract to haemolymph. It is well known from honey bees that transport of sugars from crop to midgut, as well as the rate of glucose absorption into the haemolymph, is extremely fast and matches energy requirements during flight (Crailsheim 1988a–c). Future experiments will hopefully show that Scg-AKH-II is involved in this process.

Acknowledgements. We are grateful to Özkan Alkazi for expert technical assistance and to the management of Kirstenbosch National Botanical Gardens for permission to collect carpenter bees. We thank Andrea Plos and Ian Davidson for their help with filming and processing of the videos and Heather Marco for correcting the English text. This study was partially financially supported by funds from the Foundation for Research Development and a University of Cape Town staff award.

References

Auerswald, L., Gäde, G. (1995) Energy substrates for flight in the blister beetle *Decapotoma lunata* (Meloidae). *J. Exp. Biol.* **198**:1423–1431

Auerswald, L., Schneider, P., Gäde, G. (1998) Proline powers the pre-flight warm-up in the African fruit beetle, *Pachnoda sinuata* (Cetoniinae). *J. Exp. Biol.* (in press)

Beenakkers, A.M.T., van der Horst, D.J, van Marrewijk, W.J.A. (1981) Metabolism during locust flight. *Comp. Biochem. Physiol.* **69B**:315–321

Bertsch, A. (1984) Foraging in male bumble bees (*Bombus lucorum* L.): maximising energy or minimising water load? *Oecologica* **62**:325–326

Bursell, E. (1981) The role of proline in energy metabolism. In: Downer, R.G.H. (ed) *Energy metabolism of insects.* Plenum, London, pp. 135–154

Casey, T. (1989) Oxygen consumption during flight. In: Goldsworthy, G.J, Wheeler, C.H. (eds) *Insect flight.* CRC, Boca Raton, Fla., pp. 257–272

Chappell, M.A. (1982) Temperature regulation of carpenter bees (*Xylocopa californica*) foraging in the Colorado desert of southern California. *Physiol. Zool.* **55**:267–280

Crailsheim, K. (1988a) Intestinal transport of glucose solution during honeybee flight. In: Nachtigall, W. (ed) BIONA-report 6. Akad. Wiss. Mainz Fischer, Stuttgart, pp. 119–128

Crailsheim, K. (1988b) Regulation of food passage in the intestine of the honey bee (*Apis melifera* L.). *J. Insect Physiol.* **34**:85–90
Crailsheim, K. (1988c) Intestinal transport of sugars in the honey bee (*Apis melifera* L.). *J. Insect Physiol.* **34**:839–845
Gäde, G. (1985) Isolation of the hypertrehalosaemic factors I and II from the corpus cardiacum of the Indian stick insect, *Carausius morosus*, by reversed-phase high-performance liquid chromatography, and amino acid composition of factor II. *Biol. Chem. Hoppe Seyler* **366**:195–199
Gäde, G. (1989) The hypertrehalosaemic peptides of cockroaches: a phylogenetic study. *Gen. Comp. Endocrinol.* **75**:287–300
Gäde, G. (1992) The hormonal integration of insect flight metabolism. *Zool. Jahrb. Abt. Allg. Zool. Physiol.Tiere.* **96**:211–225
Gäde, G. (1996) The revolution in insect neuropeptides illustrated by the adipokinetic hormone/red pigment-concentrating hormone family of peptides. *Z. Naturforsch.* **51c**:607–617
Gäde, G. (1997a) The explosion of structural information on insect neuropeptides. In: (Herz, W., Kirby, G.W., Moore, R.E., Steglich, W. and Tamm, C.,(eds) Progress in the Chemistry of Organic Natural Products. Springer-Verlag, Wien, pp. 1–128
Gäde, G. (1997b) Distinct sequences of AKH/RPCH family members in beetle (*Scarabaeus*-species) corpus cardiacum contain three aromatic amino acid residues. *Biochem. Biophys. Res. Commun.* **230**:16–21
Gäde, G., Auerswald, L. (1998) Insect neuropeptides regulating substrate mobilisation. *S. Afr. J. Zool* (in press)
Gäde, G., Goldsworthy, G.J., Kegel, G., Keller, R. (1984) Single step purification of locust adipokinetic hormones I and II by reversed-phase high-performance liquid chromatography and the amino-acid composition of the hormone II. *Hoppe-Seyler's Z. Physiol. Chem.* **365**:393–398
Gäde, G., Goldsworthy, G.J., Schaffer, M.H., Cook, J.C., Rinehart, K.L. Jr. (1986) Sequence analyses of adipokinetic hormones II from corpora cardiaca of *Schistocerca nitans*, *Schistocerca gregaria*, and *Locusta migratoria* by fast atom bombardment mass spectrometry. *Biochem. Biophys. Res. Commun.* **134**:723–730
Gäde, G., Hoffmann, K.-H, Spring, J.H. (1997) Hormonal regulation in insects: facts, gaps and future directions. *Physiol. Rev.* **77**:963–1032
Goldsworthy, G.J. (1983) The endocrine control of flight metabolism in locusts. In: Berridge, M.J., Treherne, J.E., Wigglesworth, V.B. (eds) *Advances in insect physiology*, vol 17. Academic, New York, pp. 149–204
Heinrich, B. (1975) Thermoregulation in bumble bees. II. Energetics of warm-up and free flight. *J. Comp. Physiol.* **96**:155–166
Louw, G.N., Nicolson, S.W. (1983) Thermal, energetic and nutritional considerations in the foraging and reproduction of the carpenter bee *Xylocopa capitata*. *J. Entomol. Soc. South Afr.* **46**:227–240

Nicolson, S.W., Louw, G.N. (1982) Simultaneous measurement of evaporative water loss, oxygen consumption and thoracic temperature during flight in a carpenter bee. *J. Exp. Biol.* **222**:287–296

Panzenböck, U., Crailsheim, K. (1997) Glycogen in Honey bee queens, workers and drones (*Apis melifera carnica* Pollm.). *J. Insect Physiol.* **43**:155–165

Rothe, U., Nachtigall, W. (1989) Flight of the honey bee IV. Respiratory quotients and metabolic rates during sitting, walking and flying. *J. Comp. Physiol.* **158B**:739–749

Sacktor, B. (1975) Utilisation of fuels by muscle. In: Candy, D.J., Kilby, B.A. (eds) *Insect biochemistry and function*. Chapman and Hall, London, pp. 1–81

Sacktor, B., Childress, C.C. (1967) Metabolism of proline in insect flight muscle and its significance in stimulating the oxidation of pyruvate. *Arch. Biochem. Biophys.* **120**:583–588

Siegert, K.J., Morgan, P.J, Mordue, W. (1985) Primary structures of locust adipokinetic hormones II. *Biol. Chem. Hoppe-Seyler* **366**:723–727

Surholt, B., Greive, H., Hommel, C., Bertsch, A. (1988) Fuel uptake, storage and use in male bumble bees *Bombus terrestris* L. *J. Comp. Physiol.* **158B**:621–629

Wheeler, C.H. (1989) Mobilisation and transport of fuels to the flight muscles. In: Goldsworthy, G.J, Wheeler, C.H. (eds) *Insect flight*. CRC Press, Boca Raton, Fla., pp. 273–303

Wilps, H., Gäde, G. (1990) Hormonal regulation of carbohydrate metabolism in the blowfly *Phormia terraenovae*. *J. Insect Physiol.* **36**:441–449

Withers, P.C. (1981) The effect of ambient air pressure on oxygen consumption of resting and hovering honey bees. *J. Comp. Physiol.* **141**:433–437

Woodring, J., Das, S., Gäde, G. (1994) Hypertrehalosaemic factors from the corpora cardiaca of the honey bee (*Apis mellifera*) and the paper wasp (*Polistes exclamans*). *J. Insect Physiol.* **40**:685–692

Zebe, E., Gäde, G. (1993) Flight metabolism in the African fruit beetle, *Pachnoda sinuata*. *J. Comp. Physiol. B* **163**:107–112

The use of a multiple antigen peptide to detect a minichromosome maintenance protein in pea (*Pisum sativum*)

**Michael S. Munns[1] · Karen Moore[2] · Lyne Jossé[2] · Paul N. Fitchett[2]
John A. Bryant[2]**

[1] Peptide and Protein Research Consultants, Washington Singer Laboratories, Perry Road, Exeter, EX4 4QG, UK
[2] School of Biological Sciences, University of Exeter, Exeter, EX4 4QG, UK

Received: 11 May 1998 / Accepted: 10 August 199890 / Published: 18 August 1998

Abstract. A 13-residue oligopeptide corresponding to a conserved region of the MCM family of proteins was synthesised as a multiple antigen peptide in which eight copies of the peptide were conjugated to an oligo-lysine core. The multiple antigen peptide was used for raising antibodies. Western blots of the polypeptides present in the DNA polymerase-α-primase complex from pea (Pisum sativum L.) were challenged with the antibodies which, even at a dilution of 1:5000, clearly revealed a polypeptide of approximately 62 kDa.

Key words. DNA polymerase · Multiple antigen peptide · MCM protein pea *Pisum sativum*

Introduction

Multiple antigen peptides

Since their introduction in 1980 (Sutcliffe et al. 1980), peptide-protein conjugates have been used widely to produce both anti-peptide and anti-protein sera which may be useful in the detection, identification and characterisation of previously unrecognised proteins. However, these conjugates exhibit a number of problems that limit their usefulness. In particular, the carrier protein itself may elicit a strong immune response, giving rise to a very high background signal. This problem may be compounded if the anti-peptide titre is low. Multiple antigen peptides were originally described by Tam (1989) as a

Correspondence to: J.A. Bryant (e-mail: J.A.Bryant@exeter.ac.uk, Fax: +44-1392-264668)

means of overcoming these problems. A multiple antigen peptide consists of multiple copies (usually eight) of a single peptide attached to a branched lysine core (Fig. 1) to give a molecule large enough to elicit a strong immune response without the interference from a large carrier protein molecule. However, despite the proven usefulness of multiple antigen peptides (McLean et al. 1991), they have not displaced peptide-protein conjugates which are still widely used. Indeed, this is first reported use of a multiple antigen peptide for study of plant proteins.

Minichromosome maintenance genes and proteins

Minichromosome maintenance (*MCM*) genes were first described, as their name implies, as genes essential for the maintenance of minichromosomes in yeast cells. *MCM* genes 2–7 form a gene family now known to be essential for viability in unicellular and multicellular eukaryotes (Kearsey et al. 1996). The proteins encoded by the *MCM* genes are involved in the initiation of DNA replication, and in a study of fission yeast it has been shown that this involves a complex of all six MCM proteins working together to „activate" the replication origins by initiating the separation of the two DNA strands (Adachi et al. 1997). For vertebrates it has been suggested that this complex forms part of the „licensing factor", ensuring that activation of replication origins (and hence initiation of replication) occurs only once per cell cycle (Madine et al. 1995). In addition, MCM3 is associated in mammalian cells with DNA polymerase-α, the polymerase involved in initiation of daughter-strand synthesis (Thömmes et al. 1992; Toyn et al. 1995) leading to our suggestion that this particular MCM protein provides a link between the activation of the origins and the initiation of the replication process per se (Bryant 1997). In plants, genes with close homology to *MCM3* have been detected in pea (*Pisum sativum*) (Moore et al. 1997) and in maize (*Zea mays*) (Sabelli et al. 1996) although nothing is known about the subcellular distribution of the corresponding protein.

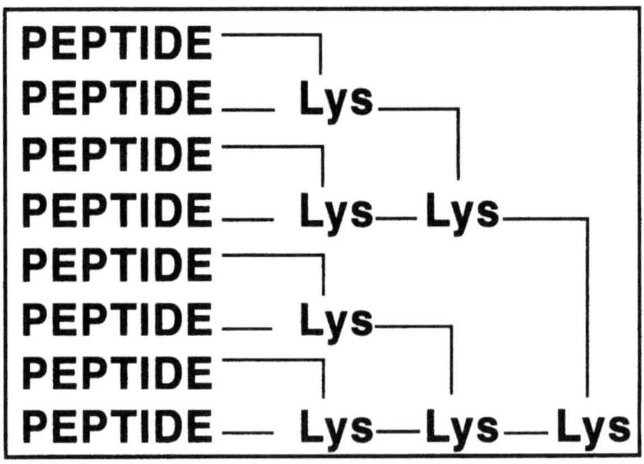

Fig. 1. Generalised structure for multiple antigen peptides

Fig. 2. Sequence of the peptide synthesised to form the multiple antigen peptide

HEVMEQQTVSIAK

Aims of this work

In the work described here the aim was to design and synthesise a multiple antigen peptide that would be used to generate antibodies for immunodetection of MCM-like proteins in pea. The results indicate that at least one MCM-like protein is associated with the pea DNA polymerase-α-primase complex.

Materials and methods

Design of the multiple antigen peptide

The peptide selected for synthesis (Fig. 2) forms part of the putative DNA-dependent ATPase domain, which is strongly conserved in all MCM proteins (Koonin 1992).

Peptide synthesis

Peptide synthesis was carried out by conventional flow techniques using di-isopropylcarbodiimide/1-hydroxybenzotriazole pre-activation of *N*-α-Fmoc-protected amino acids with acid-labile side-chain protection (Atherton and Sheppard 1989). The solid-phase resin (2.5 g, 0.1 mEq g^{-1}) was packed in a 100×50 mm glass column and reagents for synthesis were pumped through a manually operated two-valve synthesiser. The lysine core of the multiple antigen peptide was prepared by reducing the resin loading rate by 88%, adding 4-hydroxymethylphenoxyacetic acid (acid-labile linkage agent), an alanine spacer and then consecutive lysine residues (di-Fmoc-protected) until eight amino groups were available per molecule. The peptide sequence was then built up stepwise on all eight branches at the same time. Amino acids were applied in a fourfold excess in *N*,*N*-dimethylformamide (DMF) with the addition of bromophenol blue, and were recycled until the resin turned from the initial strong blue colour to yellow or pale green. After washing the resin with DMF, the Fmoc groups were cleaved by treatment with piperidine (20% in DMF) and the resin was again washed with DMF ready for the next amino acid addition. This cycle was repeated until all the amino acids were added. Final side-chain removal and cleavage from the resin were achieved simultaneously by treatment with trifluoroacetic acid:triisopropylsilane:water (95:2.5:2.5) for 2.5 h. Eluted material was concentrated and precipitated by addition of diethyl ether. The product, a white solid, was dissolved in water and freeze-dried. The purity of the product was checked by reverse-phase HPLC and by amino acid analysis.

Antibodies

Antibodies against the multiple antigen peptide were raised in rabbits using conventional techniques. This was done by Immune Systems, Paignton, Devon, UK. Briefly, rabbits were injected at monthly intervals with 0.5 mg peptide (dissolved in 1.0 ml Hunter's Titre-Max adjuvant). Bleeds were taken 10 days after each immunisation and bleeds two, three and four were used in Western blots.

Protein preparation and fractionation

DNA polymerase-α-primase was prepared from pea shoot apices as described by Bryant et al. (1992), fractionated by SDS-polyacrylamide gel electrophoresis at 22°C and blotted onto nitrocellulose as described by Al-Rashdi and Bryant (1994).

Immunodetection

Blotted proteins were probed with antibodies and the bound antibodies were detected with secondary antibodies (goat-anti-rabbit) conjugated to alkaline phosphatase as described by Fido et al. (1993).

Results and discussion

Antisera from the second, third and fourth bleeds from the immunised rabbit were used in Western blotting experiments with the DNA-polymerase-α-primase complex from pea. The second bleed did not give any clear signal but the antisera from both the third and fourth bleeds gave a very well-defined band with an apparent molecular weight of 62 kDa and much fainter bands at approximately 45 and 28 kDa (illustrated for the third bleed in Fig. 3). The 62-kDa band was clearly visible at antiserum dilutions as great as 1:5000. It was not seen in the lanes that were probed with pre-immune serum although there were two diffuse bands which reacted with the pre-immune serum (at a dilution of 1:500), one of which is visible in the lanes probed with the anti-serum.

These results show clearly the utility of multiple antigen peptides in a strategy for detecting proteins for which some sequence data are available. Particular attention is drawn to the strength and clarity of the signal obtained, even at low antiserum concentrations, and to the absence of an equivalent signal with the pre-immune serum.

Fig. 3. Western blot of the DNA polymerase-α-primase complex probed with antibodies (as described in Materials and methods) raised against the multiple antigen peptide. The third bleed from the immunised rabbit was used in this experiment. *Lane 1*: 1:1000 dilution of antiserum; *lane 2*: 1:5000 dilution of antiserum; *lane 3*: pre-immune serum, diluted 1:500; *lane 4*: molecular weight markers. In each of *lanes 1–4* the protein loading was 10 µg DNA polymerase-α-primase complex from pea (Bryant *et al.* 1992)

Equally interesting is the presence in the DNA polymerase-α-primase complex of an antigen recognised by antibodies raised against a highly conserved MCM peptide sequence. Although it has not yet been possible to obtain enough of this antigen for partial sequencing, it is suggested that it is likely to be a pea homologue of the mammalian MCM3 (sometimes known as P1) which is also associated with DNA polymerase-α (Thömmes et al. 1992). Further indirect evidence supporting this view is the identification by Moore et al. (1997), in a cDNA library from pea shoot apices, of a clone with a very marked sequence similarity to mammalian *MCM3* and to *Zmroa*, a putative *MCM3* homologue from *Zea mays* (Sabelli et al. 1996). Furthermore, the deduced amino acid sequence within the putative ATPase domain of the pea MCM3 homologue differs in only one amino acid residue from the sequence used in synthesis of the multiple antigen peptide and in two amino acid residues from the corresponding region of the human MCM3 (Fig. 4). However, it must also be pointed out that, at 62 kDa, the MCM-like protein associated with the DNA polymerase-α-primase of pea is markedly smaller than the human MCM3 protein. This may be a general feature of plant MCM proteins since the small number of such proteins detected either as antigens or as cDNAs are all smaller than those detected in other eukaryotic organisms.

Acknowledgements. We thank the Exeter University Research Fund and the BBSRC (studentship for K.M.) for financial support.

Fig. 4. Part of the putative ATPase domain in the deduced amino acid sequence of the pea MCM3 aligned with the equivalent region in the deduced amino acid sequence in human MCM3. The oligopeptide corresponding to that used for synthesis of the antigen is *underlined*

Human	AIHEVMEQGRVTIAKAGI
Pea	AIHEVMEQQTVTIAKAGI

References

Adachi, Y., Usukura, J., Yanagida, M. (1997) A globular complex formation by Nda1 and the other members of the MCM protein family in fission yeast. *Genes Cells* **2**:467–479

Al-Rashdi, J., Bryant, J.A. (1994) Purification of a DNA-binding protein from a multi-protein complex associated with DNA polymerase-α in pea (*Pisum sativum*). *J. Exp. Bot.* **45**:1867–1871

Atherton, E., Sheppard, R.C. (1989) *Solid phase peptide synthesis: a practical approach.* IRL, Oxford

Bryant, J.A. (1997) Biochemical aspects of S-phase regulation in the plant cell cycle. In: Bryant, J.A., Chiatante, D. (eds.) *Plant cell proliferation.* Wiley, Chichester, UK, pp. 31–44

Bryant, J.A., Fitchett, P.N., Hughes, S.G., Sibson, D.R. (1992) DNA polymerase-α in pea is part of a large mutiprotein complex. *J. Exp. Bot.* **43**:31–40

Fido, R.J., Tatham, A.S., Shewry, P.R. (1993) Applications of protein blotting in plant biochemistry and molecular biology. In: Bryant, J.A. (ed.) *Methods in plant biochemistry; Vol. 10, Molecular biology.* Academic, London, pp. 101–115

Kearsey, S.K., Maiorano, D., Holmes, E.C., Todorov, I.T. (1996) The role of MCM proteins in the cell cycle control of genome duplication. *Bioessays* **18**:183–190

Koonin, E.V. (1992) A common set of conserved motifs in a vast variety of putative nucleic acid-dependent ATPases including MCM proteins involved in the initiation of eukaryotic DNA replication. *Nucleic Acids Res.* **21**:2541–2547

Madine, M.A., Khoo, C.-Y., Mills, A.D., Laskey, R.A. (1995) MCM3 complex required for cell cycle regulation of DNA replication in vertebrate cells. *Nature* **375**:421–424

McLean, G.W., Owsianka, A.M., Subak-Sharpe, J.H., Marsden, H.S. (1991) Generation of antipeptide and antiprotein sera: effect of peptide presentation on immunogenicity. *J. Immunol. Methods* **137**:149–157

Moore, K.A., Aves, S.J., Bryant, J.A. (1997) Pea homologues of yeast *MCM* genes. *J. Exp. Bot. Suppl.* **48**:104

Sabelli, P.A., Burgess, S.R., Kush, A.K., Young, M.R., Shewry, P.R. (1996) cDNA cloning and characterisation of a maize homologue of the MCM proteins required for the initiation of DNA replication. *Mol. Gen. Genet.* **252**:125–136

Sutcliffe, J.G., Shinnick, T.M., Green, N., Liu, F.-T., Niman, H.L., Lerner, R.A. (1980) Chemical synthesis of a polypeptide predicted from nucleotide sequence allows detection of a new retroviral gene product. *Nature* **287**:801

Tam, J.P. (1989) High-density multiple antigen peptide system for preparation of antipeptide antibodies. *Methods Enzymol.* **168**:7–10

Thömmes, P., Fett, R., Schray, B., Burkhart, R., Barnes, M., Kennedy, C., Brown, N.C., Knippers, R. (1992) Properties of the nuclear P1 protein, a mammalian homologue of the yeast MCM3 replication protein. *Nucleic Acids Res.* **20**:1069–1074

Toyn, J.A., Toone, W.M., Morgan, B.A., Johnston, L.H. (1995) The activation of DNA replication in yeast. *Trends Biochem. Sci.* **20**:70–73

A method of direct-viewing comparative analysis of tumour cell motility in vitro; effect of pH and adhesion on cells of different malignancy

L. Pavlíková[1] · D. Zicha[2] · A. Chaloupková[1] · P. Veselý[1]

[1] Institute of Molecular Genetics, Academy of Sciences of the Czech Republic, Fleming. nam. 2, Prague, Czech Republic
[2] MRC Muscle and Cell Motility Unit, King's College London, London, UK

Received: 26 May 1998 / Accepted: 10 August 1998 / Published: 21 August 1998

Abstract. A novel method for the comparative analysis of cell motility by direct viewing was developed and used in a preliminary study of sarcoma cells. Three cell lines from the RPS family of sarcomas in inbred LEW/CUB rats which differ in the incidence of spontaneous metastasis were studied. A system of four different culture conditions, designed to mimic the stress within the tumour, was created by changing and combining two variables: the pH of the medium (with 2% calf serum only), which was either physiological at 7.4 or 6.6; and the adhesiveness of the culture substratum, which was either at a standard level or decreased. It was found that changing the pH from 7.4 to 6.6 in a single step slowed the cells down, and also changed the way in which they moved, from „walking" in random directions to moving in a more directional way. Making the culture surface less adhesive sped up cell locomotion. Decreasing the adhesiveness of the culture substratum and the pH simultaneously stimulated the migration of highly metastasizing cells preferentially. Thus we have found a way to distinguish between highly metastasizing A297Nb sarcoma cells and poorly metastasizing T15 and non-metastasizing K2 sarcoma cells, a distinction that could not be made by a simple examination of the cells. It is concluded that a comparison of cell motility by directly viewing the cells under different conditions may be useful when investigating the in vitro motility properties of malignant cells.

Key words. Motility assay · Stress conditions · Tumour cells

Correspondence to: L. Pavlíková (e-mail: luba@zeus.img.cas.cz)

Introduction

Tumour cell locomotion plays a role in tumour invasion and in the formation of metastases (Liotta and Stetler-Stevenson 1993). Therefore, the motility of neoplastic cells from experimental tumours and human malignancies may indicate their potential to cause malignancies. Although a number of indirect in vitro assays of tumour cell invasion have been devised, they often fail to predict the metastatic capacity (Noguchi et al. 1978; Noel et al. 1991). These assays are based on a histological examination that cannot, of course, analyse the cells' mode of actual motility. Therefore, it is very desirable to use time-lapse microscopy to analyse the pattern of neoplastic cell migration. We have employed computerized time-lapse grabbing (Zicha and Dunn 1995) of phase-contrast images of tumour cells cultured in a new observation chamber. The comparative assessment of motility is based on a system of four culture situations, determined by setting two variables at one of two levels – standard and stress inducing. The first variable is the pH, which is set either at a physiological level of 7.4 or slightly acid at 6.5, which simulates the extracellular pH inside the tumour (Aoki et al. 1996). The other variable is the adhesiveness, which is either standard on the tissue culture dish (T.C.) or decreased on the bacteriological polystyrene culture dishes (B.C.). The reason for choosing the less adhesive culture surface was that Veselý (1993) observed a higher incidence of the activated to locomotion morphotype in tumour cells cultured on such a substrate. The motility patterns of three neoplastic cell lines from the family of the rat RPS sarcomas (Veselý et al. 1987), which differ in their incidence of spontaneous metastasis, were studied. It was found that persistence of movement of highly metastasizing A297Nb cells, which were not shown to disperse significantly more than poorly metastasizing cells (Pavlíková et al., in preparation) in the indirect QTCM test (Matoušková et al. 1989), is induced by conditions of stress. The results achieved justify the method, which involves challenging the cells in vitro by stress-inducing culture conditions (i.e. slightly acidic medium and less adhesive substratum), as it reveals the otherwise hidden motility potential of the highly metastasizing sarcoma cells.

Materials and methods

Cells

Three neoplastic cell lines were used, all of which belong to the family of spontaneously metastasizing RPS sarcomas in inbred LEWIS rats (Veselý et al. 1987). They differ in the actual incidence of spontaneous metastasis (Pavlíková et al. in preparation).

The K2 IX clonal cell line is a neoplastic population derived from in vitro spontaneously transformed embryonic fibroblasts of the inbred rat strain

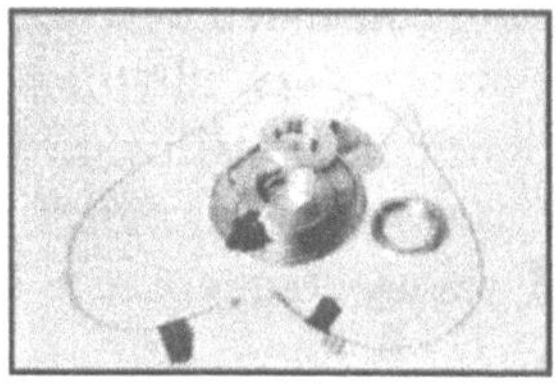

Fig. 1. Temperature-controlled observation chamber CTOC 35. Its design enables the use of standard T.C. 35-mm polystyrene culture dishes and the medium to be changed during filming without disturbance

LEW/CUB (Veselý and Weiss 1973). The subpopulation IX was chosen because it does not metastasize spontaneously after subcutaneous injection into inbred LEW/CUB rats.

The T15 IV clonal cell line was derived from the K2 cells, after going through neoplastic progression in vitro and in vivo, by clonal selection for morphotype and malignancy (Veselý et al. 1987). The T15 IV cells induced subcutaneous tumours that shed metastases in 25% of the inoculated animals in the lung.

The A297 Nb II cell line was isolated in a similar way to the T15 cells. The A297Nb cells repeatedly caused a 100% incidence of spontaneous metastasis in the lung.

Culture media

M1H medium was used for culturing the cells. This medium is Eagle's minimum essential medium with all non-essential amino acids supplemented with 10% calf serum.

F2 medium is the same medium, but with the $NaHCO_3$ concentration decreased to one-third (0.35g%), without phenol red, supplemented with 2% calf serum only, and set to the appropriate pH level, i.e. 7.4 or 6.6, using 20 mM TES or BES buffer respectively.

Microscope and image analysis equipment

A Nikon Diaphot 300 inverted microscope with LWD phase-contrast optics, a high-performance COHU CCD camera and a Sony Trinitron Monitor, all attached to Matrox grabber in Cyrix 166 MHz PC, were used. The programs dr.exe for time-lapse grabbing of images and dp.exe for projecting the images and evaluation of *x,y* co-ordinates of cell positions were originally developed by D. Zicha (Zicha and Dunn 1995). The Microsoft Excel 97 program was used for evaluation of the data obtained.

A new design for a temperature-controlled observation chamber (CTOC 35)

The new chamber (Fig. 1) was made to accommodate either a standard 35-mm polystyrene T.C. dish or dish-like holder of 24-mm-diameter round coverslips for high-resolution work. It has provision for exchanging media and the in-built controlled heating regulates the temperature of the cells under observation at 37°C while the background temperature of the whole microscope is maintained just below 30°C by a separate temperature-control system. The CTOC 35 was designed and made in collaboration with the Special Optical Systems, Prague, Czech Republic.

Preparation of cells for time-course microscopy

Cells were seeded on 35-mm diameter T.C. dishes (NUNC, Roskilde, Denmark) in M1H medium. After a 48-h cultivation, the dish was washed once with F2 medium at the appropriate pH. Then the same medium was added and the dish was placed in the heated CTOC 35 chamber, which was then placed in the microscope for preliminary stabilization of the environment. When cells were to be cultivated on the bacteriological surface (B.C.), they were seeded onto a 60-mm-diameter dish on the internal side of an inserted 33-mm-diameter disc cut out from the bottom of the bacteriological dish (GAMA, České Budějovice, Czech Republic). After a 48-h cultivation in M1H medium, the disc was taken out, washed in F2 medium at the appropriate pH, put into a 35-mm-diameter NUNC cultivation dish with the same medium, and placed in the heated CTOC 35 chamber. Then after a 1-h period of stabilization in the microscope the time-lapse recording of images of selected suitable cells began.

Design of the experiment

Every experiment lasted for 6 h. The interval between pictures was 2 min. The resulting movie (series of digital images) contains 180 pictures each of which indicates x,y positions of approximately ten cells in one frame selected for the minimum of cell-to-cell contacts.

Assessment of cell motility

$$TCT = \sum_{i=1}^{N-1} T_i$$

A tracking program (Zicha and Dunn 1995) was used to acquire positions (x_i,y_i) of a cell over N frames (i=1,...,N). For a cell remaining in the observation field during the whole film N=180. The total cell track (TCT) was calculated as a summary of distances between cell positions, where:

$$T_i = \sqrt{(x_{i+1} - x_i)^2 + (y_{i+1} - y_i)^2}$$

TCT is expressed in μm. The actual cell speed (ACS_i) over each 2-min interval was calculated as $T_i/2$ min and is therefore expressed in μm min^{-1}. The mean cell speed (MCS) of the cell was calculated as a mean value of ACS:

$$MCS = \frac{\sum_{i=1}^{N-1} ACS_i}{N-1}$$

Mean population speed (MPS) was calculated as average MCS from all cells in a film. For final appreciation, the idealized cell track (ICT) for each experiment was computed by multiplying the MPS by 360 min.

Results

Analysis of data

A slightly acidic medium pH of 6.6 compared with the standard pH of 7.4 on the T.C. surface caused a drop of the ICT of all cells examined to almost half. Raw data are shown in Tables 1 and 2 and the change in the percentage of the standard value (pH 7.4 and T.C. surface) in Table 3.

A slightly acid medium pH of 6.6 (compared with the standard pH of 7.4) on the B.C. surface caused a drop of the ICT of all cells examined, but to a slightly lesser extent than on the T.C. surface; for K2 cells the change was almost negligible. The speeds are shown in Tables 1 and 2 and the percentage change in the standard value (pH 7.4 and T.C. surface) in Table 3.

The less adhesive surface (compared with the T.C. surface) at both pH values tested caused the ICT of A297Nb cells to increase significantly and the ICT of T15 cells to increase markedly, but it had no significant effect on that of K2 cells. Raw data are shown in Tables 1 and 2 and the percentage change in the standard value (pH 7.4 and T.C. surface) in Table 3.

Table 1. Review of the idealized cell tracks (*ICTs*) under the combinations of the variables examined. The values are given in μm. The *ICT* is computed by multiplying the average speed of the population under the given conditions (*MCS*) by total length of the experiment (360 min)

	pH 6.6		pH 7.4	
	T.C.	B.C.	T.C.	B.C.
K2IX	56.08	72.00	90.73	85.80
T15IV	139.00	263.30	278.50	404.90
A297NbII	139.50	437.13	242.70	731.20

Table 2. Review of the mean speed of the cell population (*MPS*) in the combinations of the variables examined. The values are given in μm/min. *MPS* is mean value of mean speed of all cells (*MCS*) on the film. The B.C./T.C. column shows how much more is the *MPS* on bacteriological surface higher than on the tissue surface

	pH 6.6			pH 7.4		
	T.C.	B.C.	B.C./T.C.	T.C.	B.C.	B.C./T.C.
K2 IX	0.157	0.201	1.28	0.253	0.240	0.94
T15 IV	0.388	0.735	1.89	0.778	1.131	1.45
A297NbII	0.389	1.187	2.57	0.678	2.040	3.01

Table 3. The relative changes of ICT (it is the same as for MPS). The data in the *first two rows* are the ICT values at pH 6.6 as a percentage of the *ICT* value at pH 7.4 for cells grown on the tissue culture dish (*TCD*) and the bacteriological culture dish (*BCD*). The data in the *third row are the* ICT values for cells grown on BCDas a percentage of the ICT values for cells grown on the standard TCD at pH 7.4; data in the *fourth row* are as for the *third row*, but for cells grown at a pH of 6.6

		K2IX	T15IV	A297NbII
pH 6.6/pH 7.4	TCD	63%	50%	57%
	BCD	84%	65%	60%
B.C./T.C.	pH 7.4	94%	145%	301%
	pH 6.6	128%	189%	257%

Analysis of images

The nature of the cell tracks is shown in Figs. 2–4 (parts a–d). It is clear that changing the pH to 6.6 increased the persistence of cell locomotion on the standard T.C. surface, with the effect being least significant on the K2 cells and most significant on the A297Nb cells, with that on the T15 cells being intermediate. This effect is further enhanced by the bacteriological surface which by itself stimulates the motility at both values of pH.

Figure 5 shows histograms of the distribution of the motile activity of the five most motile A297NbII cells under the following conditions: (1) pH 7.4 and T.C. surface; (2) pH 6.6 and T.C. surface; (3) pH 7.4 and B.C. surface; (4) pH 6.6 and B.C. surface. The ACS_i (where i=1,…,180 for every cell) values are given on

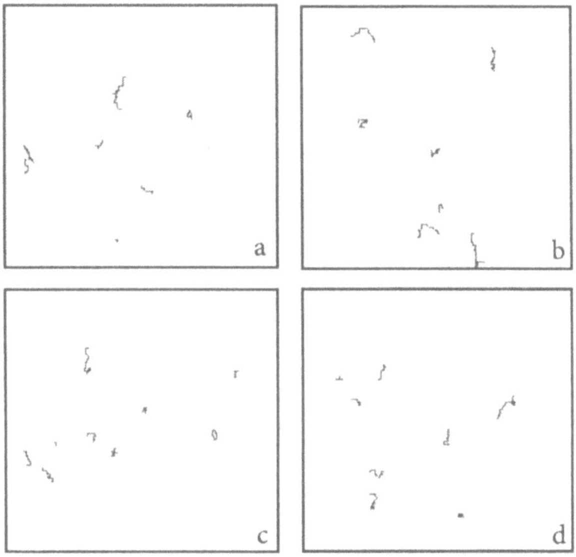

Fig. 2a–d. Cell trajectories of all K2 IX cells from the films taken under the following experimental conditions. **a** pH 7.4 and tissue culture surface; **b** pH 6.6 and tissue culture surface; **c** pH 7.4 and bacteriological surface; **d** pH 6.6 and bacteriological surface. The total length of the cell's trajectory was designated TCT

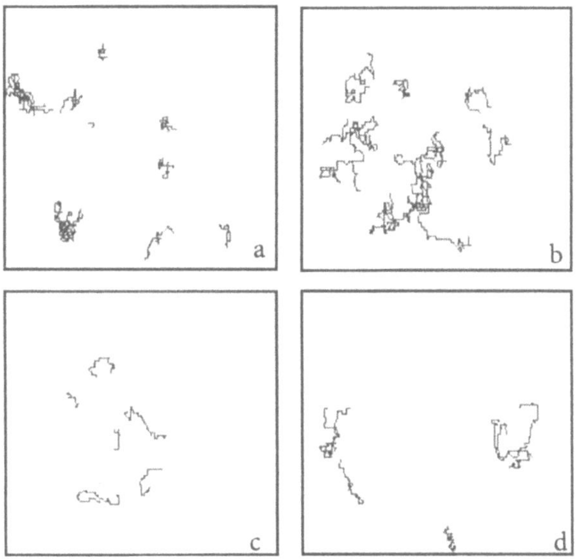

Fig. 3a–d. Cell trajectories of all T15 IV cells from the films taken under the following experimental conditions. **a** pH 7.4 and tissue culture surface; **b** pH 6.6 and tissue culture surface; **c** pH 7.4 and bacteriological surface; **d** pH 6.6 and bacteriological surface. The total length of the cell's trajectory was designated TCT

the graph. The frequency of movement speed in the population is on *y*-axis. These results show the role of changing the pH and the surface adhesiveness in the induction of the accelerated motility of the most malignant cells examined.

Figures 6–8 (parts a–d) show the initial (time-zero) image from the movies obtained in the experiments. Notice the changed phenotype of cells on the bacteriological dish surface in parts b and d of the figures. The demonstration movies are shortened versions of the original recordings produced by taking every fifth picture only.

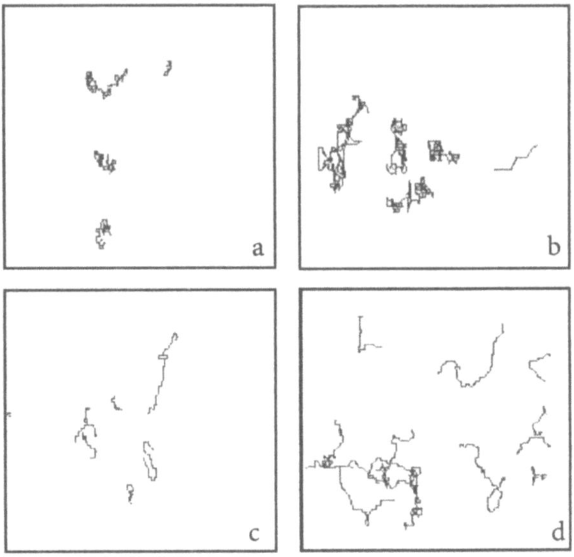

Fig. 4a–d. Cell trajectories of all A297Nb II cells from the films taken under the following experimental conditions. **a** pH 7.4 and tissue culture surface; **b** pH 6.6 and tissue culture surface; **c** pH 7.4 and bacteriological surface; **d** pH 6.6 and bacteriological surface. The total length of the cell's trajectory was designated TCT

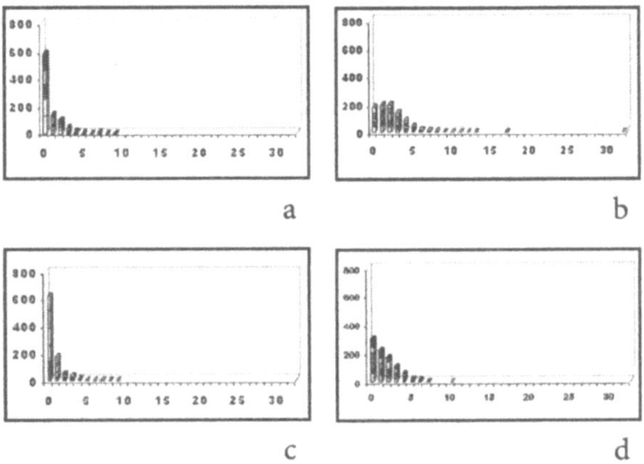

Fig. 5a–d. Histograms showing the distribution of A297NbII cell motile activity under various conditions. The five most motile cells from every movie recorded were chosen. Their actual values of speed during all 2-min intervals (*ACS*) observed are given in the *graph*. The ACS values throughout the experiment are given on the x-axis in μm/min and the frequency of the occurrence is given on y-axis. The cells were investigated under the following conditions: **a** pH 7.4 and tissue culture dish; **b** pH 6.6 and tissue culture dish; **c** pH 7.4 and bacteriological tissue dish; **d** pH 6.6 and bacteriological tissue dish

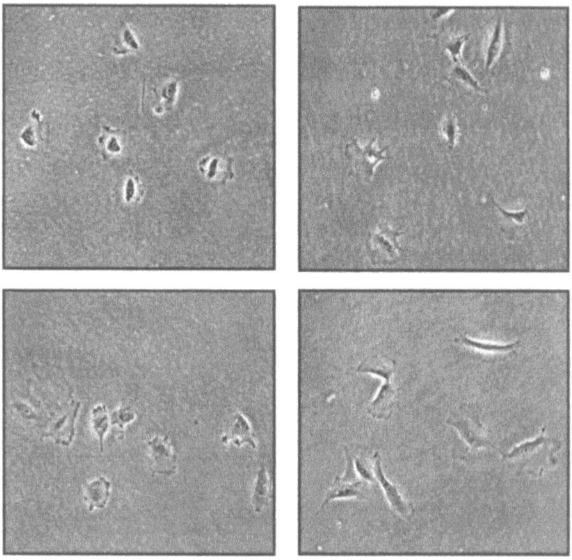

Fig. 6. First images of the movies showing the K2 IX cell line at: **a** pH 7.4 and tissue culture surface; **b** pH 7.4 and bacteriological culture surface; **c** pH 6.6 and tissue culture surface; **d** pH 6.6 and bacteriological culture surface. The shorten version of each movie demonstrating the behaviour of the cells under specified conditions can be seen by *clicking* on this *first picture*

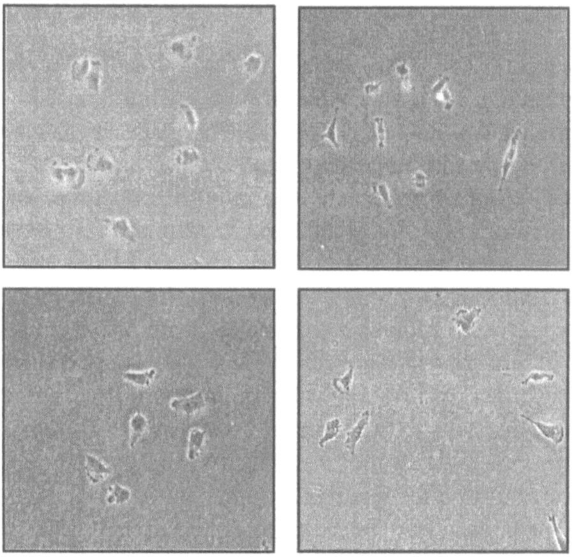

Fig. 7. First images of the movies showing the T15 IV cell line at: **a** pH 7.4 and tissue culture surface; **b** pH 7.4 and bacteriological culture surface; **c** pH 6.6 and tissue culture surface; **d** pH 6.6 and bacteriological culture surface. The shorten version of each movie demonstrating the behaviour of the cells under specified conditions can be seen by *clicking* on this first *picture*

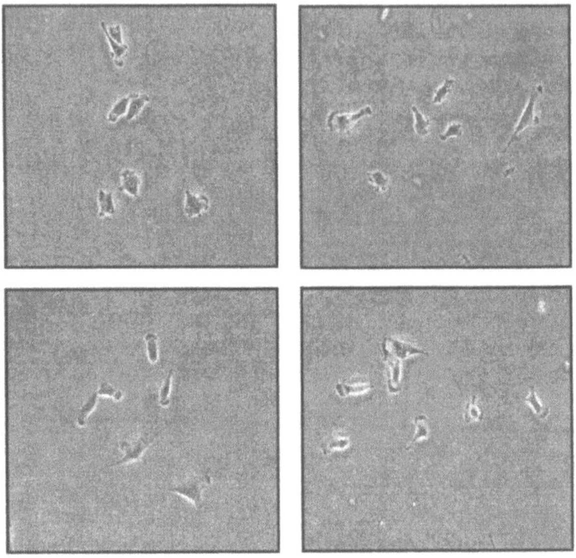

Fig. 8. First images of the movies showing the A297Nb II cell line at: **a** pH 7.4 and tissue culture surface; **b** pH 7.4 and bacteriological culture surface; **c** pH 6.6 and tissue culture surface; **d** pH 6.6 and bacteriological culture surface. The shorten version of the movie demonstrating the behaviour of the cells under specified conditions can be seen by *clicking* on this first *picture*

Discussion

Tumour cell invasion and metastasis are the major cause of death in cancer patients. They are both complex phenomena. The course of the invasion and metastasis has been divided by investigators into a series of defined steps each of which can be further analysed. As the locomotory potential of malignant cell plays an important role in both invasion as well as metastasis, there have been numerous attempts to characterize it in in vivo (McDonald et al. 1998) and in in vitro systems.

Although a number of in vitro assays of tumour invasion have been developed (e.g. chick chorioallantoic membrane invasion, mouse lung invasion, chemotaxis-invasion assay, invasion to the monolayer of fibroblasts, chick embryonic skin invasion) they fail often to predict the metastatic capacity of the tumour cells (Noguchi et al. 1978; Noel et al. 1991). The few indirect assays of tumour cell locomotive capacity rarely reveal the metastatic potential of the tumour in question. For example, in the in vitro wound colonization assay, i.e. determination of speed of migration of cells into the artificial wound made in the confluent sheet of cells (Liao et al. 1995), it is very difficult to determine the contribution of neat cell migration versus pressure from behind caused by proliferation of the surrounding cells. The QTCM test developed in our laboratory (Matoušková et al. 1989) indirectly estimates the migration of cells at the population level by comparative evaluation of the dispersion of cells in colonies derived from single cells after 72 h of cultivation. The results obtained imply that cells are more motile in acidic media (pH 6.6), as the dispersion at this pH was higher for every cell population examined. In the majority of the cell lines examined, the increase of the final dispersion coincided with the increase of their in vivo metastatic activity. The QTCM test could not differentiate between A297Nb cells, which repeatedly had a 100% incidence of spontaneous metastasis in inbred rats LEW/CUB, and T15 cells, which have less metastatic activity (L. Pavlíková, unpublished observation). Therefore, we decided that direct observation of cell motility may provide more relevant information with respect to the metastatic potential of these cell types.

We have devised an assay based on comparing the motile behaviour of cells under different conditions in vitro. It was hoped at the outset that such an approach, designed to mimic the in vivo conditions in the tumour, would yield more valuable information about the locomotory potential of the cells under study. After a thorough consideration of the importance, reliability and manageable complexity of the experimental methods, we choose two factors – the pH and the adhesion of the culture surface. In order to imitate intratumour conditions, a slightly acid pH of 6.6 and a less adhesive surface of the bacteriological dishes were used to challenge the cells. For the selection of an acid pH we took our previous works into consideration. This simulation of intratumour conditions (Aoki et al. 1996) led to increased directionality of movement (Veselý et al. 1987; Zicha et al. 1993), greater secretion of cathepsin B (Křepela et al. 1989), an altered actin cytoskeleton (Pokorná et al. 1994), increased intracellular motility (Veselý et al. 1994) and increased secretion of neutral proteinases (Chaloupka et al. 1998) in our model rat sarcoma cell lines. The occurrence of the activated to locomotory phenotype (Veselý 1993) of the rat sarcoma cells cultured on the bacteriological dish was the main reason for using a culture dish with a less adhesive surface. We used medium containing 2% calf serum only, as this was found to be best for investigating chemotaxis in vertebrate cells (D. Zicha, unpublished observation). A lower serum concentration seems to be a factor that slightly reduces cell multiplication and activates movement.

The results obtained by directly viewing the cells when in an acidic pH contradict those obtained using the QTCM method (Matoušková et al. 1989) and time-lapse cinemicroscopy of related rat RPS sarcoma cell lines (Zicha et al. 1993; Pokorná et al. 1994). In all our experiments presented in this article, the surface speed of the cells at pH 7.4 was higher than at 6.5. In the QTCM test, the dispersion of the colony is attributed mainly to cell multiplication and the increased directionality of cells in the G1 phase of growth. The eventual regressive movement of the cell is not reflected by the final intercellular distances. It is not even reflected by the position of the cells measured every 20 min, which was the case in the time-lapse data of experiments of Zicha et al. (1993) and Pokorná et al. (1994). It accounts for the fallacious presumption that cells have greater motility at an acidic pH, which was derived from the results of the QTCM test which only measures the final position of the cells. The long sampling period used in the cinemicroscopy method (i.e. 20 min between frames) means that this technique could not resolve the random nature of cell motility at a pH of 7.4. The more accurate direct-viewing approach with ten times shorter intervals between frames (2 min versus 20 min) reveals the random but faster walk at pH 7.4 and the persistence of cell migration at pH 6.6.

The bacteriological polystyrene B.C. surface caused the cells to accelerate. While the more malignant cell lines were able to accelerate by a factor of 2–3, the least malignant K2 IX cell line was not sensitive to the stimulation provided by the B.C. surface, and their movement remained constant and slow. At this stage of the study it would be premature to speculate about the role of adhesion molecules connected to the cytoskeleton in this phenomenon or about the greater production of autocrine motility factor on the less adhesive surface. Nevertheless the difference between the mode and extent of migration of the neoplastic cell lines examined correlates with their metastatic potential in vivo.

Thus it can be deduced that the ability of highly malignant cells to cope in vivo with both an acidic intercellular pH and impaired adhesion can provide highly malignant cells with the opportunity to migrate, which is the beginning of the metastatic cascade. It seems justifiable to conclude that by comparing the pattern of cell motility under different conditions by directly viewing the cells, we are able to detect locomotory attributes of neoplastic cells that are significant for malignancy.

References

Aoki, Y., Akagi, K., Tanaka, Y., Kawai, J., Tokahashi, M. (1996) Measurement of intratumor pH by pH indicator used in ^{19}F magnetic resonance spectroscopy. Measurement of extracellular pH decrease caused by hyperthermia combined with hydralazine. *Invest. Radiol.* **31**:680–389

Chaloupka, J., Kučerová, H., Váchová, L., Krchňáková, E., Chaloupková, A., Pavlíková, L., Zicha, D., Veselý, P. (1998) Effect of pH on proteinase secretion by transformed fibroblast populations. *Folia Biol. (Prague)* **44**:111–116

Dunn, G.A., Zicha, D. (1993) Long-term chemotaxis of neutrophils in stable gradients: preliminary evidence of periodic behavior. *Blood Cells* **19**:25–41

Křepela, E., Veselý, P., Chaloupková, A., Zicha, D., Urbanec, P., Rasnick, D., Vičar, J. (1989) Cathepsin B in cells of two rat sarcomas with different rates of spontaneous metastasis. *Neoplasma* **36**:529–540

Liao, G., Nagasaki, T., Gundersen, G. (1995) Low concentrations of nocodazole interfere with fibroblast locomotion without significantly affecting microtubule level: implications for the role of dynamic microtubules in cell locomotion. *J. Cell. Sci.* **108**:3473–3483

Liotta, L.A., Stetler-Stevenson, W.G. (1993) Principles of molecular cell biology of cancer: cancer metastasis. In: DeVita, V.T. Jr., Hellmen, S., Rosenberg, S.A. (eds) *Cancer: principles and practice of oncology*, 4th edn. Lippincott, Philadelphia, N.Y., pp 134–149

Matoušková, E. Zicha, D. Urbanec, P. Veselý, P. (1989) Quasi-dynamic test of in vitro cell migration developed to characterize dividing cells of spontaneously metastasizing rat sarcomas. *Folia Biol. (Prague)* **36**:117–129

McDonald, I.C., Schmidt, E.E., Morris, C.L., Groom, A.C., Chambers, A.F. (1998) In vivo videomicroscopy of experimental hematogenous metastasis: cancer cell arrest, extravasation and migration. In: Soll, D.R., Wessels, D. (eds) *Motion analysis of living cells.* Wiley-Liss, New York, pp. 263–288

Noel, A.C., Calle, A., Emonard, H.P., Nusgens, B.V., Simar, L., Foidart, J., Lapiere, C.M., Foidart, J.M. (1991) Invasion of reconstituted basement membrane matrix is not correlated to the malignant metastatic cell phenotype. *Cancer Res.* **51**:405–414

Noguchi, P.D., Johnson, J.B., O'Donnel, R., Petricciani, J.C. (1978) Chick embryonic skin as a rapid organ culture assay for cellular neoplasia. *Science* **199**:980–982

Pokorná, E., Jordan, P.W., O'Neill, C.H., Zicha, D., Gilbert, C.S., Veselý, P. (1994) Actin cytoskeleton and motility in rat sarcoma cell populations with different metastatic potential. *Cell Motil. Cytoskeleton* **28**:25–33

Veselý, P. (1993) Phase contrast microscopy in the in vitro study of neoplastic cell dynamic morphology and behavior as related to malignancy. In: Pluta, M., Szyjer, M. (eds) Proceedings of SPIE: phase contrast and differential interference contrast imaging techniques and applications, Vol. 1846, pp. 140–145. The Society of Photo-Optical Instrumentation Engineers.

Veselý, P., Weiss, R. (1973) Cell locomotion and contact inhibition of normal and neoplastic rat cells. *Int. J. Cancer* **11**:64–76

Veselý, P., Chaloupková, A., Urbanec, P., Urbancová, H., Boháč, L., Krchňáková, E., Franc, F., Šprincl, L., Vousden, K., Moss, R., Heaysmann, J., Dilly, P.N. (1987) Patterns of in vitro behaviour characterizing cells of spontaneously metastasizing K2M rat sarcoma. *Folia Biol. (Prague)* **33**:307–324

Veselý, P., Luers, H., Riehle, M., Bereiter-Hahn, J. (1994) Substraction scanning acoustic microscopy reveals motility domains in cells in vitro. *Cell Motil. Cytoskeleton* **29**:231–240

Zicha, D., Dunn, G.A. (1995) An image processing system for cell behaviour studies in subconfluent cultures. *J. Microsc.* **179**:11–21

Zicha, D., Pokorná. E., Urbanec, P., Chaloupková, A., Veselý, P. (1993) Acid pH induced persistent motility in vitro of metastasizing sarcoma cells. *J. Comput. Assist. Microsc.* **5**:273–279

The influence of hibernation patterns on the critical enzymes of lipogenesis and lipolysis in prairie dogs

Craig L. Frank[1] · **Stephen P.J. Brooks**[2] · **Henry J. Harlow**[3] · **Kenneth B. Storey**[4]

[1] The Louis Calder Center, Biological Station and Department of Biological Sciences, Fordham University, PO Box K, Armonk, NY 10504, USA
[2] Nutrition Research Division, HPB, Health Canada, 3W Banting Research Centre, Tunney's Pasture 2203C, Ottawa, Ontario K1A 0L2, Canada
[3] Department of Zoology and Physiology, University of Wyoming, Laramie, WY 82071, USA
[4] Department of Biology, Carleton University, Ottawa, Ontario K1S 5B6, Canada

Received: 19 May 1998 / Accepted: 2 September 1998 / Published: 14 September 1998

Abstract. White-tailed prairie dogs (*Cynomys leucurus*) are spontaneous hibernators that enter torpor each fall, whereas black-tailed prairie dogs (*C. ludovicianus*) hibernate facultatively only when food- or water-stressed during the winter. The body masses of both species greatly increase during the fall feeding period, with most of this gain in the form of depot fat. Body fat is utilized during winter fasting and/or hibernation. We measured the activities of fatty acid synthase (FAS), ATP-citrate lyase (ACL), malic enzyme (ME), glucose-6-phosphate dehydrogenase (G6PDH), and hormone-sensitive lipase (HSL) in the tissues of both *C. leucurus* (hibernating and euthermic) and *C. ludovicianus* (euthermic only) under controlled conditions. The activities of FAS, ACL, and G6PDH in the liver all decreased during hibernation. The activities of ME and G6PDH in white adipose tissue (WAT) were also reduced during hibernation. Euthermic *C. leucurus* and euthermic *C. ludovicianus* differed only in brown adipose (BAT) ACL and WAT G6PDH activities. No significant differences in HSL activities were found between these two species or between euthermic and hibernating animals. These results suggest that this seasonal body fat cycle is due, at least in part, to seasonal variations in the activities of FAS, ME, ACL, and G6PDH that affect the rate of fatty acid synthesis. This study also demonstrates that spontaneous hibernators do not have a greater capacity to synthesize fatty acids during the fall than facultative hibernators, as previously suggested.

Correspondence to: C.L. Frank (e-mail: frank@murray.fordham.edu, Fax: +1-914-2732167)

Key words. Adipose · ATP citrate lyase (ACL) · *Cynomys* · Glucose-6-phosphate dehydrogenase (G6PDH) · Hibernation · Hormone-sensitive lipase (HSL) · Liver · Malic enzyme (ME)

Introduction

Ground squirrels (*Spermophilus* sp.), marmots (*Marmota* sp.), and prairie dogs (*Cynomys* sp.) are all closely related members of the same monophyletic group in the family Sciuridae (Hafner 1984), and they hibernate for periods of up to 8 months during winter (Heaney 1984). These hibernators are often divided into two broad categories: spontaneous and facultative. Spontaneous hibernators enter torpor every fall/winter regardless of the amount of food and water available. Facultative hibernators, in contrast, remain euthermic throughout most of the fall and winter, and become torpid only when food and/or water is unavailable (Folk 1974; Davis 1976; Mrosovsky 1978). Previous field studies reveal that, during late summer/early fall, the body masses of both spontaneous and facultative hibernators increase by 35–45%, mostly in the form of body fat (Morton 1975; Kenagy and Barnes 1988; Thompson et al. 1993). Feeding rates more than double during this period (Loehr and Risser 1977; Kenagy 1987). Although depot fats are the main (but not only) source of energy utilized during torpor by spontaneous hibernators (Kayser 1965), facultative hibernators metabolize large amounts of protein as well as fat during torpor (Harlow 1995). Since the body masses (and fat contents) of all hibernators decrease dramatically during winter (Kenagy and Barnes 1988; Thompson et al. 1993), hibernation is considered to be a biochemical state comparable to prolonged fasting (Krilowicz 1985).

The biochemical basis of this pronounced annual cycle of body fat content is poorly understood. The short-term regulation of fatty acid synthesis is mediated by the reversible phosphorylation of acetyl CoA carboxylase (Szepesi et al. 1989), whereas its long-term regulation is due to changes in the activities of the enzymes of fatty acid synthesis (Mead et al. 1986). Long-term regulation is achieved by varying cellular mRNA contents for enzymes such as glucose 6-phosphate dehydrogenase (G6PDH), malic enzyme (ME), ATP-citrate lyase (ACL), and fatty acid synthase (FAS) which in turn regulates the amount of these enzymes synthesized (Wakil et al. 1983). The regulation of fatty acid oxidation in mammals is complex, but a key step in this process is the control of hormone-sensitive lipase (HSL) by reversible phosphorylation. HSL is the primary enzyme responsible for the liberation of fatty acids from stored triacylglycerols (Mead et al. 1986), and it is activated through phosphorylation by protein kinase A.

Laboratory studies conducted by Mostafa et al. (1993) involving tissues from euthermic and hibernating marmots (*Marmota flaviventris*) concluded that the FAS activity in both liver and white adipose tissue (WAT) decreases dramatically during hibernation. The examination of liver tissues from both euthermic and hibernating golden-mantled ground squirrels (*Spermophilus lateralis*) by Walter et al. (1996) also suggested that the activity of FAS in liver decreases during hibernation. Studies of free-ranging Richardson's ground squirrels (*S. richardsonii*), a spontaneous hibernator, revealed much greater FAS activity in WAT during the fall fattening period compared with the FAS activity in the WAT of free-ranging black-tailed prairie dogs (*C. ludovicianus*), a facultative hibernator (Turner et al. 1989; Thompson et al. 1993). These results led Thompson et al. (1993) to propose that spontaneous hibernators generally have a greater capacity to synthesize fatty acids than facultative hibernators. A fourfold increase in HSL mRNA was found in the WAT of hibernating marmots (*M. flaviventris*) when compared to that of WAT from euthermic marmots, thus it has been suggested that lipolysis increases during hibernation (Wilson et al. 1992).

The results of these studies conducted by Turner et al. (1989), Mostafa et al. (1993), Wilson et al. (1992), Thompson et al. (1993), and Walter et al. (1996) are all impossible to interpret because of the following flaws in their designs. FAS activity in liver and WAT increases as the total lipid content of the diet decreases (Herzberg 1983; Mead et al. 1986). Laboratory rats maintained on a diet with a high fructose/starch ratio have higher FAS, ACL, ME, and G6PDH activities in their livers than those maintained on a diet containing a relatively lower fructose/starch ratio (Fiebig et al. 1998). These studies indicate that all experiments investigating the enzymes involved in fatty acid synthesis must consider the lipid content and fructose/starch ratio of the diet. All of the tissues examined in the studies conducted by Turner et al. (1989) and Thompson et al. (1993), and those of the euthermic group in Walter et al. (1996), were from free-ranging animals. Diet composition was unknown in these studies, thus it is quite possible that some or all of the differences in enzyme activities observed were caused by dietary variations in total lipid content and/or fructose/starch ratios. The „hibernating“ tissues examined by Mostafa et al. (1993) were in fact collected from euthermic marmots that had been previously forced to arouse from torpor; therefore, it is not certain that the enzyme activities reported by these authors actually represent those during natural torpor. Wilson et al. (1992) did not measure the actual activities of phosphorylated HSL in the marmot WAT samples that they examined, thus their hypothesis that lipolysis increases during hibernation lacks support.

What is the relationship between hibernation and (1) fatty acid synthesis, or, (2) lipolysis? Do spontaneous hibernators have a greater capacity to synthesize fatty acids than facultative hibernators? We conducted laboratory experiments with white-tailed (*C. leucurus*) and black-tailed (*C. ludovicianus*) prairie dogs in order to better understand the biochemical basis of the body fat cycles associated with hibernation. White-tailed prairie dogs are spontaneous hibernators, whereas black-tailed prairie dogs are facultative hibernators (Harlow and Menkens 1986). We measured the tissue activities of four enzymes associated with fatty acid synthesis (G6PDH, ME, ACL, and FAS) and a fatty acid oxidation enzyme (HSL) in both of these species.

Materials and methods

Twelve adult white-tailed prairie dogs (*C. leucurus*) and six adult black-tailed prairie dogs (*C. ludovicianus*) were captured in southeastern Wyoming during the summer of 1993. These rodents were housed individually in large cages at the University of Wyoming and maintained at an ambient temperature (T_a) of 22°C on a natural (fall) photoperiod. The plant parts ingested by free-ranging sciurids have lipid contents ranging from 7 to 30% (Eshelman and Jenkins 1989; Frank et al. 1998). All animals were thus fed a modified version of Purina 5001 rodent diet with a fat content of 15%. This diet also contained 21% protein, 49% carbohydrate, 7% starch, 5% fiber, and 3% minerals. This diet was the only food type given to the prairie dogs during the study. After 37 days of feeding and fattening (on 12 October 1993), eight *C. leucurus* and all *C. ludovicianus* were sacrificed. Tissue samples were immediately removed from each carcass, frozen in liquid N_2, and stored at -80°C. The remaining four *C. leucurus* were allowed to continue feeding and fattening for an additional 36 days. At the end of this period (on 7 November 1993), these prairie dogs (in their cages) were placed at 5.0°C in an environmental chamber to induce hibernation. Within 7 days, all four animals began hibernating and they were kept under these conditions for 10 weeks. They were then sacrificed while torpid and their tissues were stored at -80°C. All euthermic animals were sacrificed by decapitation after first being anesthetized with CO_2 gas. Torpid *C. leucurus* were sacrificed by decapitation after first being placed in CO_2 gas for 1 min, thus all animals were treated in the same manner.

The activities of FAS, ME, ACL, and G6PDH were determined in crude homogenates prepared by grinding frozen tissue (1:4) in 50 mmol/l imidazole, 5 mmol/l EDTA, 5 mmol/l EGTA, 100 mmol/l NaF, and 30 mmol/l beta-mercaptoethanol (pH=7.0), and 0.1 mM phenylmethylsulfonyl fluoride. These homogenates were centrifuged for 15 min at 12,000 *g*. Enzyme activities were measured spectrophotometrically at 25°C. FAS was measured according to the technique of Nepokroeff et al. (1975), ACL activity was measured using the method of Takeda et al. (1969), ME activity was determined according to Hsu and Lardy (1969), and G6PDH activity was determined as described by Deutsch (1983). Total HSL activity was measured using a modified version of the techniques described by Sheridan and Harmon (1994). This assay does not phosphorylate (activate) HSL, it only measures the activity of the phosphorylated HSL already present in the sample. Total tissue protein contents were determined using the Biuret method after homogenization in 5% sodium dodecylsulfate plus 0.5 N NaOH and incubation for 1 h at 30°C (Brooks et al. 1995). Enzyme activities were measured in the liver, WAT, and brown adipose tissue (BAT). All enzyme activities were calculated in nmol $min^{-1}mg^{-1}$ tissue protein in order to factor out the effects of seasonal changes in tissue lipid content. The enzyme activities of the three experimental groups were compared using the general linear models ANOVA procedure (SAS Institute 1985).

Results

The liver FAS activities of euthermic white-tailed prairie dogs and euthermic black-tailed prairie dogs were equivalent (Fig. 1), and both were significantly greater than those of hibernating white-tailed prairie dogs (F=3.46, P=0.04). These three groups did not significantly differ in FAS levels found in either their WAT (F=0.11, P=0.90) or BAT (F=1.26, P=0.31; Fig. 1).

The liver ACL levels in euthermic *C. leucurus* and *C. ludovicianus* were also statistically equivalent (Fig. 2), but both were significantly greater than that of hibernating *C. leucurus* (F=3.81, P=0.04). Euthermic *C. leucurus*, *C. ludovicianus*, and hibernating *C. leucurus* did not significantly differ in the ACL activities found in their WAT (F=0.22, P=0.80), although euthermic *C. leucurus* had significantly greater ACL activity in their BAT (F=6.72, P=0.008) than the other two groups (Fig. 2).

Euthermic *C. leucurus*, *C. ludovicianus*, and torpid *C. leucurus* all had statistically equivalent ME activities (Fig. 3) in both their liver (F=2.92, P=0.08) and BAT (F=3.32, P=0.06). The WAT of hibernating *C. leucurus* had a significantly lower ME activity than the WAT from the other groups (F=3.97, P=0.04; Fig. 3).

Hibernating white-tailed prairie dogs had liver G6PDH activities (Fig. 4) that were less than half of those found in the euthermic groups (F=4.58, P=0.03). The WAT from euthermic white-tailed prairie dogs had more than twice the G6PDH activity (Fig. 4) than that of either other group (F=16.96, P=0.0001). Euthermic *C. leucurus*, *C. ludovicianus*, and torpid *C. leucurus* did not significantly differ in BAT G6PDH activities (F=0.74, P=0.50; Fig. 4), and

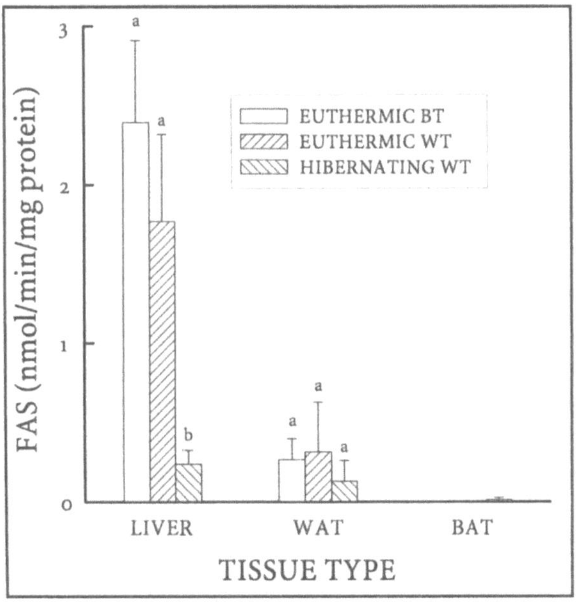

Fig. 1. Mean (±SE) fatty acid synthase (*FAS*) specific activities in tissues from white-tailed (*WT*) and black-tailed (*BT*) prairie dogs. For each euthermic BT, euthermic WT, and hibernating WT mean, n=6, 8, and 4, respectively. Means for the same tissue labeled with the same *lower case letter* are not significantly different at the P<0.05 level

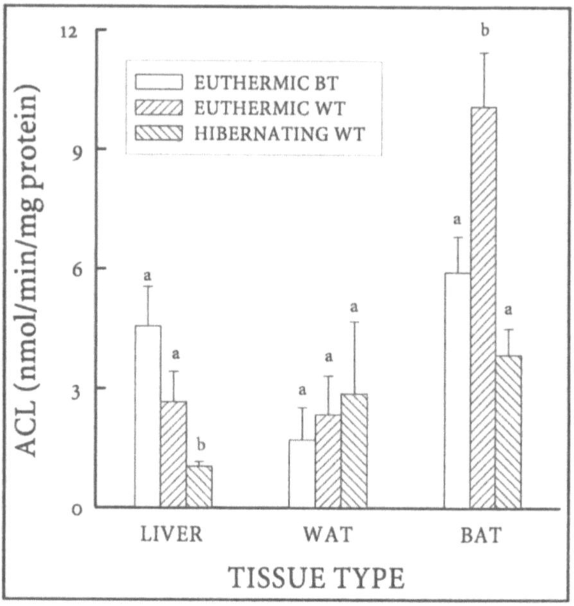

Fig. 2. Mean (±SE) ATP-citrate lyase (*ACL*) specific activities in tissues from white-tailed (*WT*) and black-tailed (*BT*) prairie dogs. Sample sizes and symbols are the same as those for Fig. 1

neither did they significantly differ in the activity of HSL (Fig. 5) in either their WAT (F=0.73, P=0.50) or BAT (F=0.66, P=0.53).

Discussion

Hibernating *C. leucurus* had significantly lower liver FAS, liver ACL, BAT ACL, WAT ME, liver G6PDH, and WAT G6PDH activities than euthermic *C. leucurus*. The results of this study thus clearly support the hypothesis that fatty acid synthesis decreases during hibernation. The synthesis of fatty acid molecules in the cytosol is carried out by FAS. The dramatic decrease in liver FAS activity during hibernation thus probably reduces the overall rate of fatty acid synthesis. The synthesis of one fatty acid molecule by FAS requires one molecule of acetyl CoA and at least 14 molecules of reduced nicotinamide adenine dinucleotide phosphate (NADPH). The required acetyl CoA is released into the cytosol by ACL, whereas the large amounts of NADPH needed for fatty acid synthesis are produced by G6PDH and ME (Gumma et al. 1969; Mead et al. 1986). The results of this study thus suggest that fatty acid synthesis during hibernation my also be inhibited by a reduction in the level of acetyl CoA and the rate of NADPH production in the cytosol.

A number of studies have demonstrated that the overall rate of fatty acid synthesis is greatest during the fall fattening period. Liver slices from 13 lined ground squirrels (*S. tridecemlineatus*) had a twofold higher rate of fatty acid synthesis in the fall than during the summer (Whitten and Klain 1969). The

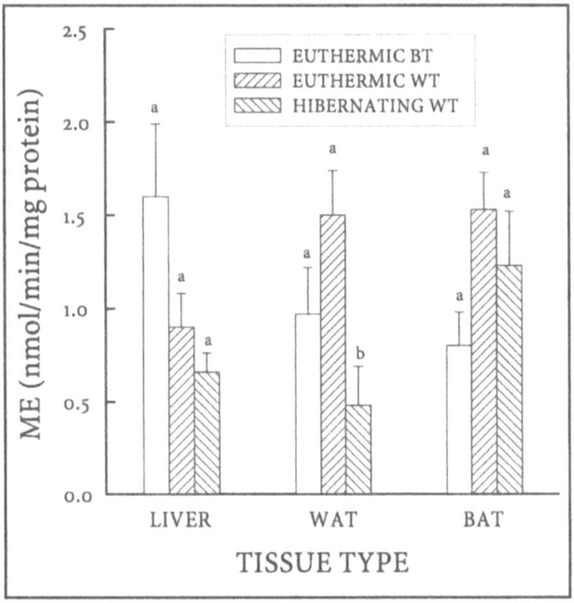

Fig. 3. Mean (±SE) malic enzyme (*ME*) specific activities in tissues from white-tailed (*WT*) and black-tailed (*BT*) prairie dogs. Sample sizes and symbols are the same as those for Fig. 1

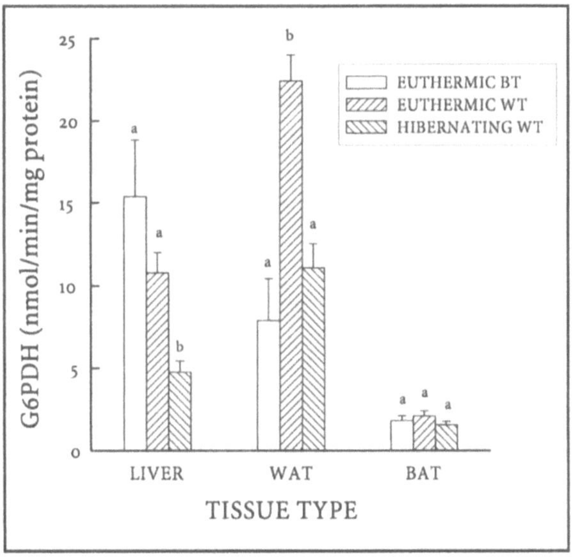

Fig. 4. Mean (±SE) glucose-6-phosphate dehydrogenase (*G6PDH*) specific activities in tissues from white-tailed (*WT*) and black-tailed prairie (*BT*) prairie dogs. Samples sizes and symbols are the same as those for Fig. 1

WAT of Syrian hamsters has a fivefold increased ability to synthesize lipids in vitro after cold exposure (Baumber and Denyes 1963). The WAT of free-ranging and euthermic *C. ludovicianus* has a much greater FAS activity during the fall than during winter (Thompson et al. 1993). The results of our study, when interpreted in conjunction with these studies, suggest that the increase in fatty acid synthesis during the fall is due, at least in part, to increased activities of FAS, ACL, ME, and G6PDH.

The changes observed in both liver and WAT enzyme activities between the prehibernatory fattening and hibernation periods indicate that both of these tissues participate in seasonal fatty acid synthesis. The relative contribution of each of these tissues to lipogenesis by hibernating mammals is poorly understood. During fattening by *S. tridecemlineatus*, the liver appears to be the major site of fatty acid synthesis (Whitten and Klain 1969). When Syrian hamsters were cold exposed, however, the ability of their WAT to synthesize lipids in vitro increased substantially, whereas the in vitro lipogenesis in liver slices from these animals decreased (Denyes and Carter 1961; Denyes and Baumber 1964). The relative contributions of liver and WAT in prehibernatory lipogenesis may vary with species, and thus requires further investigation.

Only the ACL activity of BAT, and the G6PDH activity of the WAT from euthermic *C. leucurus* were significantly greater than the levels of these enzymes found in the corresponding euthermic tissues of *C. ludovicianus*. In all other cases the enzyme activities of these two euthermic groups were equivalent. These findings contradict the hypothesis that spontaneous hibernators have a greater capacity to synthesize fatty acids during the fall fattening period than facultative hibernators, as proposed by Thompson et al. (1993). Our study indicates instead that these two hibernator groups have roughly the same fatty acid synthesis capacity during this period. We did not find any significant differences in adipose tissue HSL activities among the three groups. These results are consistent with those of Florant et al. (1993), who observed that the rate of glycerol release by WAT (an indication of lipolysis) is identical in both hibernating and euthermic marmots (*M. flaviventris*). This is surprising in view of the increase in HSL mRNA levels observed by Wilson et al. (1992) in WAT during hibernation, and it does not support their hypothesis that lipolysis in-

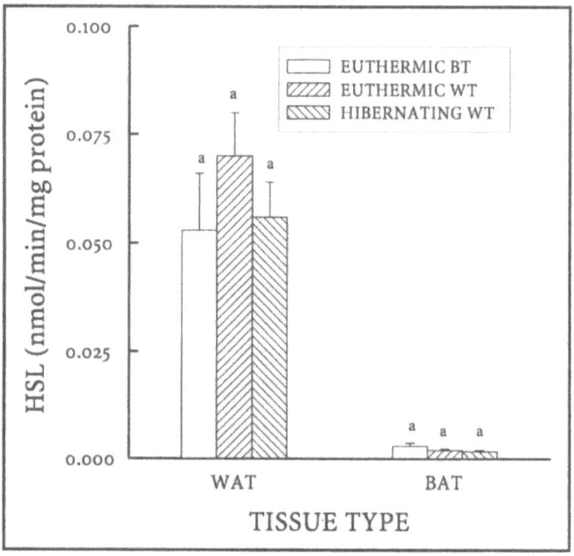

Fig. 5. Mean (±SE) specific hormone-sensitive lipase (*HSL*) activities in tissues from white-tailed (*WT*) and black-tailed (*BT*) prairie dogs. Sample sizes and symbols are the same as those for Fig. 1

creases during hibernation. Our findings demonstrate instead that the levels of active (phosphorylated) HSL present in both WAT and BAT during hibernation are the same as those during the fall feeding/fattening period. It appears that the regulation of HSL may be through both translation and phosphorylation.

The nature of the metabolic signal that produces the seasonal changes in the activities of FAS, ACL, G6PDH, and ME observed is unknown. The mechanism that keeps the level of HSL activity in the WAT constant despite a fourfold increase in HSL mRNA levels is also unknown. These aspects of the seasonal body fat cycles found in hibernating mammals warrant additional study. Further investigation of this system will not only provide new insights into mammalian hibernation, but also the mechanism of body fat regulation in general.

Acknowledgements. This study was supported in part by a grant from the North Atlantic Treaty Organization awarded in 1992 to C.L. Frank, and a grant from the Natural Sciences and Engineering Research Council of Canada awarded to K.B. Storey.

References

Baumber, J., Denyes, A. (1963) Acetate-1-^{14}C metabolism of white fat from hamsters in cold exposure and hibernation. *Am. J. Physiol.* **205**:905–911

Brooks, S.P.J., Lampi, B.J., Sewar, G., Botting, H.G. (1995) A comparison of methods for determining total body protein. *Anal. Biochem.* **226**:26–30

Davis, D.E. (1976) Hibernation and circadian rhythms of food consumption in marmots and ground squirrels. *Q. Rev. Biol.* **51**:477–514

Denyes, A., Baumber, J. (1964) Lipogenesis of cold-exposed and hibernating golden hamsters. *Ann. Acad. Sci. Fenn.* **71**:129–138

Denyes, A., Carter, J.D. (1961) Utilization of acetate-1-^{14}C by hepatic tissue from cold exposed and hibernating hamsters. *Am. J. Physiol.* **205**:1043–1048

Deutsch, J. (1983) Glucose 6-phosphate dehydrogenase. In: Bergmeyer, H.U. (ed.) Methods of enzymatic analysis. Verlag Chemie, Weinheim, Germany, pp. 190–197

Eshelman, B.D., Jenkins, S.H. (1989). Food selection by Belding's ground squirrels in relation to plant nutritional features. *J. Mammal.* **70**:846–851

Fiebig, R., Griffiths, M.A., Gore, M.T., Baker, D.A., Oscai, L., Ney, D.M., Ji, L.L. (1998) Exercise training down-regulates hepatic lipogenic enzymes in meal-fed rats: fructose versus complex-carbohydrate diets. *J. Nutr.* **128**:810–817

Florant, G.L., Foster, J., Ram, P.K., Edens, N. (1993) Seasonal lipolysis in hibernators. In: Carey, C., Florant, G.L., Wunder, B.A., Horwitz, B. (eds.) *Life in the cold.* Westview, Boulder, pp. 225–232

Folk, E.G. (1974) Environmental physiology. Lea and Febiger, Philadelphia
Frank, C.L., Dierenfeld, E.S., Storey, K.B. (1998) The relationship between lipid peroxidation, hibernation, and food selection in mammals. *Am. Zool.* **38**:341–349
Gumma, K.H., Novello, F., McLean, P. (1969) The pentose phosphate pathway of glucose metabolism. Hormonal and dietary control of the oxidative and nonoxidative reactions and related enzymes of the cycle in adipose tissue. *Biochem. J.* **114**:253–265
Hafner, D.J. (1984) Evolutionary relationships of the Nearctic Sciurudae. In: Murie, J.O., Michener, G.R. (eds.) *The biology of ground-dwelling squirrels.* University of Nebraska Press, Lincoln, pp. 3–23
Harlow, H.J. (1995) Fasting biochemistry of representative spontaneous and facultative hibernators: the white-tailed prairie dog and the black-tailed prairie dog. *Physiol. Zool.* **68**:915–934
Harlow, H.J., Menkens G.E. Jr (1986) A comparison of hibernation in the black-tailed prairie dog, white-tailed prairie dog, and Wyoming ground squirrel. *Can. J. Zool.* **64**:793–796
Heaney, L.R. (1984) Climatic influences on life history tactics and behavior of North American tree squirrels. In: Murie, J.O., Michener, G.R. (eds.) *The biology of ground-dwelling squirrels.* University of Nebraska Press, Lincoln, pp. 43–78
Herzberg, G.R. (1983) The influence of dietary fatty acid composition on lipogenesis. In: Draper, H.H. (ed.) *Advances in nutritional research*, vol 5. Plenum, New York, pp. 221–253
Hsu, R.Y., Lardy, H.A. (1969) Malic enzyme. *Methods Enzymol.* **13**:230–235
Kayser, C. (1965) Hibernation. In: Mayer, W.V., Van Gelder, R.G. (eds.) *Physiological mammalogy*, vol II. Academic, New York, pp. 179–279
Kenagy, G.J. (1987) Energy allocation for reproduction in the golden-mantled ground squirrel. *Symp. Zool. Soc. Lond.* **57**:259–273
Kenagy, G.J., Barnes, B.M. (1988) Seasonal reproductive patterns in four coexisting rodent species from the Cascade Mountains, Washington. *J. Mammal.* **69**:274–292
Krilowicz, B.L. (1985) Ketone body metabolism in a ground squirrel during hibernation. *Am. J. Physiol.* **249**:R462–R470
Loehr, K., Risser, A.C. (1977) Daily and seasonal activity patterns of the Belding ground squirrel in the Sierra Nevada. *J. Mammal.* **58**:445–448
Mead, J., Alfin-Slater, D., Howton, D., Popjak, G. (1986) *Lipids: chemistry, biochemistry, and nutrition.* Plenum, New York
Morton, M.L. (1975) Seasonal cycles of body weights and lipids in Belding ground squirrels. *Bull. S. C. Acad. Sci.* **74**:128–143
Mostafa, N., Everett, D.C., Chou, S.C., Kong, P.A., Florant, G.L., Coleman, R.A. (1993) Seasonal changes in critical enzymes of lipogenesis and triacylglycerol synthesis in the marmot (*Marmota flaviventris*). *J. Comp. Physiol. B* **163**:463–469

Mrosovsky, N. (1978) Circannual cycles in hibernators. In: Wang, L.C.H., Hudson, J.W. (eds.) *Strategies in cold: natural torpidity and thermogenesis.* Academic, New York, pp. 21–65

Nepokroeff, C.M., Lakshmanan, M.R., Porter, J.W. (1975) Fatty acid synthase from rat liver. *Methods Enzymol.* **35**:37–44

SAS Institute. 1985. *SAS user's guide: statistics*, version 5. SAS Institute, Cary, N.C.

Sheridan, M.A., Harmon, J.S. (1994) Adipose tissue. In: Hockachka, P.W., Mommsen, T. P. (eds) *Biochemistry and molecular biology of fishes*, vol 3. Elsevier, Amsterdam, pp. 305–311

Szepesi, B., Kamara, A.K., Clarke, S.D. (1989) Lack of specificity of polyunsaturated fats in the inhibition of rat liver glucose 6-phosphate dehydrogenase. *J. Nutr.* **119**:161–165

Takeda, Y., Suzuki, F., Inoue, H. (1969) ATP-citrate lyase (citrate cleavage enzyme). *Methods Enzymol.* **13**:152–160

Thompson, T.A., Agar, M.W., Bintz, G.L. (1993) Lipid deposition and use by black-tailed prairie dogs *Cynomys ludovicianus* in the natural environment. *Physiol. Zool.* **66**:561–579

Turner, J.J., Andersen, B.O., Bintz, G.L. (1989) Whole-body lipids and fatty acid synthetase activity in Richardson's ground squirrel, *Spermophilus richardsonii. Physiol. Zool.* **62**:1383–1397

Wakil, S.J., Stoops, J.K., Joshi, V.C. (1983) Fatty acid synthesis and its regulation. *Annu. Rev. Biochem.* **52**:537–565

Walter, R.D., Wang, P., Bhat, B.G., Igal, R.A., Florant, G.L., Coleman, R.A. (1996). Hepatic enzymes of glycerolipid synthesis: comparison of seasonal changes in the golden-mantled ground squirrel (*Spermophilus lateralis*) and the yellow-bellied marmot (*Marmota flaviventris*). In: Geiser, F., Hulbert, A.J., Nicol, S.J. (eds) *Adaptations to the cold.* University of New England Press, Armidale, pp. 233–236

Whitten, B.K., Klain, G.J. (1969) NADP-specific dehydrogenases and hepatic lipogenesis in the hibernator. *Comp. Biochem. Physiol.* **29**:1099–1105

Wilson, B.E., Deeb, S., Florant, G.L. (1992) Seasonal changes in hormone-sensitive and lipoprotein lipase mRNA concentrations in marmot whole adipose tissue. *Am. J. Physiol.* **262**:R177–R181

The ecophysiological significance of lung-air retention during submersion by the air-breathing crabs *Cardisoma carnifex* and *Cardisoma hirtipes*

Stephen Morris · Jocelyn Dela-Cruz

School of Biological Sciences (A08), University of Sydney, NSW 2006, Australia

Received: 6 August 1998 / Accepted: 28 October 1998 / Published: 6 November 1998

Abstract. The air-breathing crabs *Cardisoma hirtipes* and *C. carnifex* trap 18.8 and 15.5 ml of air respectively within the lumen of the lung (branchial chamber) when submerging, rather than breathe water, and suffer a hypoxic hypercapnia. The role, effect and advantage of trapping air within the lung during submergence was assessed for both species. *Cardisoma* retained at least 85% of the lung air during a 30 min submergence. The mass-specific O_2 uptake rate (MO_2) of air-breathing *C. hirtipes* and *C. carnifex* was reduced by at least 60% during submersion and the crabs additionally consumed 80% or more of the O_2 stored in the trapped air. After 30 min of submergence the large arterial–venous difference in haemolymph partial pressure of O_2 (PO_2) and O_2 content (CO_2) of air-breathing *Cardisoma* was completely removed, consistent with near-zero O_2 transport. Removing the air from the lung lumen did not restore MO_2 but rather deprived the crab of an O_2 store and, in *C. hirtipes*, promoted anaerobiosis. Submergence reduced the haemolymph CO_2 by 50% or more, regardless of the presence of trapped air in their lungs. *C. hirtipes* and *C. carnifex* with a retained air bubble lost Na at only 2.6 and 2.3 μmol $g^{-1} h^{-1}$, respectively, but at 3.9 and 5.0 μmol $g^{-1} h^{-1}$ without the air bubble. Unidirectional Na uptake in *C. hirtipes* was only 0.90 μmol $g^{-1} h^{-1}$ when air was trapped in the lung but 1.74 μmol $g^{-1} h^{-1}$ when the bubble was removed. In *C. carnifex* these rates were 1.88 and 2.79 μmol $g^{-1} h^{-1}$ respectively. *C. hirtipes* and *C. carnifex* both trap air within the lung and avoid exposing exchange surfaces to water. There is no large respiratory advantage to expelling the air but there are significant ion regulation cost savings in retaining it.

Key words. Air-breathing · Amphibious · Carbon dioxide · *Cardisoma* Crustacean · Haemolymph · Land crabs · Na regulation · Oxygen · Terrestrial

Correspondence to: S. Morris (e-mail: stevems@bio.usyd.edu.au)

Introduction

Land crabs of the genus *Cardisoma* have a distribution that shows they retain a clear dependency on water (Gifford, 1962; Hartnoll, 1988) since their burrows often descend to the local water table (Wood and Boutilier, 1985; Pinder and Smits, 1993; Adamczewska and Morris, 1996; Morris and Adamczewska, 1996). In these crabs the branchiostegal linings of the branchial chambers that hold the gills have become elaborated into a functional lung (Farrelly and Greenaway, 1992, 1993). It has been suggested that *Cardisoma* may be bimodal breathers and use water to facilitate excretion of CO_2 across their gills (e.g. Wood and Randall, 1981; Farrelly and Greenaway, 1994), or that they submerge and breathe water to different extents (e.g. Gifford, 1962; Shah and Herreid, 1978; Cameron, 1981; O'Mahoney and Full, 1984; Burggren *et al.*, 1985) because of the need to avoid or compensate for dehydration in air. However, the flooded burrows of *Cardisoma* are generally hypoxic and hypercapnic (Wood and Boutilier, 1985; Pinder and Smits, 1993; Adamczewska and Morris, 1996) and would not facilitate respiratory gas exchange.

Some species of *Cardisoma* have been observed to trap air bubbles within the lung lumen (branchial chamber) and this behaviour appears to be the normal submergence response of *C. hirtipes* (Adamczewska and Morris, 1996) and *C. carnifex* (Morris and Adamczewska, 1996). Cameron (1981) observed that similar air bubbles had to be removed from *C. carnifex* if they were to survive submersion. However, recent investigations of responses of *Cardisoma*, with trapped air, to submersion showed that the respiratory gas exchange rate of *C. hirtipes* in water is less than half that in air (Dela-Cruz and Morris, 1997a). Consequently, *C. hirtipes* and *C. carnifex* submerged *in situ* became internally hypoxic compared to when they breathed air (Adamczewska and Morris, 1996; Morris and Adamczewska, 1996; Dela-Cruz and Morris, 1997a) which they do for the majority of the time (Dela-Cruz and Morris, 1997b). Similarly, the submergence of *C. carnifex* (Morris and Adamczewska, 1996) or of *C. hirtipes* (Adamczewska and Morris, 1996; Dela-Cruz and Morris, 1997a) failed to induce the respiratory alkalosis indicative of gas exchange with water. The effect of submergence on haemolymph partial pressure of CO_2 (PCO_2) may very well be dependent on the extent to which the gills are bathed. Adamczewska and Morris (1996) suggest that the air bubble may provide an oxygen store and prevent diffusive ion losses to freshwater. However, in addition to long-term maintenance of hydration state, periodic access to water is required for nitrogen (NH_3/NH_4^+) excretion. In *C. hirtipes* (Dela-Cruz and Morris, 1997b) and *C. carnifex* (Wood and Boutilier, 1985) nitrogen is stored and excreted in rapid, intense bursts during brief immersion episodes, as found also for the freshwater crab *Potamonautes warreni* (Morris and van Aardt, 1998), limiting the distribution away from water.

The apparent aversion of *Cardisoma* to immersion remains puzzling. *C. carnifex* have, however, a greater haemolymph osmotic pressure (by 200 mosmol/l) than *C. hirtipes*, seemingly sustained by periodic visits to seawater (Morris and Adamczewska, 1996). In contrast, *Cardisoma* lose ions when immersed in freshwater (e.g. Wood and Boutilier, 1985; Dela-Cruz *et al.*, submitted), creating an extra expense for regulation, and this may be a reason for minimizing submergence. This problem is exacerbated in *C. hirtipes* which has routine access to only freshwater.

The gills of *Cardisoma* may generally be important in ion regulation (Greenaway, 1989; Wolcott, D.L., 1991; Wolcott T.G., 1992). The gills of *C. hirtipes* contain a high-affinity Na uptake system (Greenaway, 1989); *C. hirtipes*, *C. carnifex* and *C. guanhumi* both survive water with a Na concentration as low as 0.5 mmol l^{-1} (Herreid and Gifford, 1963; Greenaway, 1989; Dela-Cruz *et al.*, submitted) but at a considerable greater cost for *C. carnifex* compared with *C. hirtipes* (Dela-Cruz *et al.* submitted) while *C. armatum* is also a hyper-regulator (De Leersnyder and Hoestlandt, 1963). However, some *Cardisoma* at least can manage ion balance without frequent access to open water by using the gills to reprocess urine and reabsorb salts from the urine as it passes over the gills (Dela-Cruz and Morris, 1997b). During submersion some branchial surfaces must be bathed to facilitate the removal of NH_3/NH_4^+ (Dela-Cruz and Morris, 1997b; Morris and van Aardt, 1998). The retention of an air bubble within the lung lumen (branchial chamber) may serve during such short-term submergence to minimize ion efflux. If so, this constitutes a potential behavioural regulation of salt balance and attendant costs, but may limit the surfaces available for respiratory gas exchange with water.

The possibility that retaining air in the lung lumen, rather than ventilating with water, might compromise respiratory gas exchange and transport but minimize ion regulation costs while submerged was investigated in *C. hirtipes* and the more marine *C. carnifex*. Submerged crabs were compared to air-breathing crabs, and in addition the effect of removing the air bubble during submergence was assessed. Respiratory, acid-base and metabolic responses were determined as were the maximum Na loss rate and the unidirectional Na uptake rate in crabs with and without air bubbles trapped within the lung lumen. Specifically the importance of the trapped air in determining O_2 uptake and respiratory gas transport was assessed against the possibility that flooding of the branchial/lung surfaces causes an expensive or even unsustainable ion loss, in turn promoting the adoption of a terrestrial life-style in the *Cardisoma.*

Materials and methods

Gecarcinid crabs, *Cardisoma hirtipes* (Dana) (371–457 g) and *C. carnifex* (Herbst) (362–462 g) were collected from Christmas Island (Parks Australia permit) and West Island of the Cocos (Keeling) Atoll. The location of these islands in the Indian Ocean, and the situation of the collection sites are described elsewhere (Adamczewska and Morris, 1996; Morris and Adamczewska, 1996). Briefly, *C. hirtipes* were collected around inland freshwater springs and *C. carnifex* adjacent to the central lagoon of the Cocos atoll. The crabs were packed in shredded paper and transported (AQIS[1] permit) by aircraft to Sydney. Both species were housed at 25±2°C (the temperature for all experiments) and 85% relative humidity in communal terrariums which contained 2 l water containers that were sufficient to allow the crabs to immerse. Routinely, the water provided was either artificial Christmas Island Water (CIW; Adamczewska and Morris, 1996) for *C. hirtipes* or 10% seawater for *C. carnifex*. The crabs were fed weekly on mixed fruit and dry dog food. The light cycle was 14 h light:10 h dark and all experiments were performed during the light phase.

Effective lung volume and gas composition

The volume of air trapped as a bubble within the lung (or simplicity the vascularized branchiostegal lining of the branchial chamber, which also houses the gills, is referred to as the lung and the space within as the lung lumen) during submersion of *C. hirtipes* and *C. carnifex* was determined by a displacement method (Dela-Cruz and Morris, 1997a). A 15 cm-diameter funnel was attached to a 50 ml measuring cylinder with a water-tight seal and the whole system flooded with water by submergence in a plastic tub. Resting crabs were individually submerged beneath the funnel that collected trapped air into the cylinder. The amount of air was recorded after 5 min and 30 min of submergence. Subsequently, the crab was inverted and agitated vigorously to release all the remaining air in the lung (branchial chambers) into the cylinder.

The PO_2 and PCO_2 within the lung air bubble was determined in both *C. hirtipes* and *C. carnifex* either while air-breathing or subsequent to 30 min of submergence. To sample the lung lumen gas the crabs were bilaterally fitted with short polyethylene catheters which penetrated from the apex of the dorsal surface through the carapace just into the air-space of the lung, as described by Morris and van Aardt (1998). These catheters had the added function of ensuring that the branchial chambers/lungs were flooded when required, by manually withdrawing all air with a syringe. To sample lung gas at least 5 ml of air was withdrawn into a glass syringe and introduced into serially connected D616 thermostated cells containing either an O_2 or a CO_2 electrode (Radiometer Pacific), and was chased by a second 5 ml sample taken from the other lung. There was never any fluid in either sample. Partial pressures were recorded by a PHM 73 meter (Radiometer) as kPa. The CO_2 electrode was calibrated using humidified precision gas mixtures (Linde Australia, Sydney) and the O_2 using O_2-free gas and air-saturated water.

[1] AQIS – Australian Quarantine Inspection Service

Oxygen uptake rates

The rate of O_2 uptake for air-breathing crabs was determined by flow-through respirometry using an Applied Electrochemistry O_2 analyser (Morris *et al.*, 1994; Dela-Cruz and Morris, 1997a). The values for *C. hirtipes* are also reported in Dela-Cruz and Morris (1997a) but are compared here with those for *C. carnifex*.

The rate of O_2 uptake for *C. hirtipes* and *C. carnifex* submerged with or without the lung air was determined using a closed respirometry method (Dela-Cruz and Morris, 1997a). Briefly, an individual crab was placed in a cylindrical respirometer and submerged in aerated water for 30 min. Water sampled immediately prior to and after the 30 min was passed over a Radiometer electrode. Rates (MO_2) were calculated as μmol g^{-1} h^{-1} for both *Cardisoma* species. All crabs were fitted with catheters as described above. No attempt was made to determine if there were any small differences in the activity of crabs in air or water since this is integral to the exposure treatments.

Haemolymph measurements

The haemolymph contents and partial pressures of O_2 and CO_2 in *C. hirtipes* and in *C. carnifex* that were either breathing air or submerged for 30 min (CIW and 30% SW respectively), with or without a trapped air bubble within the lung lumen (n=6 for each species), were measured. The carapace of the crabs was drilled a minimum of 24 h prior to experiments to allow haemolymph sampling from the pericardial cavity and from the efferent pulmonary vessel (Farrelly and Greenaway, 1993). Each crab was housed in an individual box with a supply of fresh drinking water for 24 h prior to experimentation. Each crab was sampled for pulmonary (post pulmonary vessel, content and pH only), arterial (pericardial cavity) and venous haemolymph (from the venous sinuses at the base of the last walking leg). The haemolymph samples were immediately analysed for partial pressure and content of O_2 and CO_2, as well as haemolymph pH, using a BMS 3 MK II Blood Micro System thermostated at 25±0.2°C and connected to a PHM73 pH/Blood gas monitor (Radiometer, Copenhagen, Denmark). The electrodes were calibrated daily with humidified gases. The O_2 electrode was calibrated using O_2-free gas and air-saturated water. The CO_2 electrode was calibrated using two humidified gases, one with 0.5% and the other 2.5% CO_2 (precision CIG certified analysis). The pH electrode was calibrated with Radiometer precision buffers of pH 7.410 (S1510) and 6.865 (S1500), accurate to ±0.005 at 25°C.

Haemolymph O_2 contents were measured using the modified Tucker chamber method (Tucker, 1967) as outlined by Bridges *et al.* (1979). The O_2 electrode was maintained at 32°C and connected to an oxygen meter (Strathkelvin model 781). The changes in PO_2 were recorded on a pen recorder (Kipp and Zonen, model BD111). The haemolymph CO_2 content was measured using a Corning 965 analyser (calibrated with HCO_3^- standard, 15 mmol l^{-1}).

An aliquot of the remaining haemolymph samples was mixed (ratio 1:1) with 0.6 mol l^{-1} $HClO_4$ to denature proteins and neutralized with 2.5 mol l^{-1} K_2CO_3. The denatured sample was centrifuged at 10,000 *g* for 10 min and the supernatant was used for l-lactate analysis (Boehringer Mannheim test kit no. 138 084). Whole haemolymph samples were also analysed for glucose (Sigma Diagnostics test-kit no. 510) and urate concentrations (Sigma Diagnostics test kit no. 685).

Na flux measurements

To determine if the lung air minimized ion loss from the crabs to extra-corporeal water the maximum rate of Na loss was determined in both *C. hirtipes* and *C. carnifex* which had been pre-acclimated for 3 weeks to drinking and immersion in 30% seawater. This elevated haemolymph Na concentration, which was similar in all *C. hirtipes* with a mean 288.9±1.4 mmol l^{-1} and averaged 331.8±3.0 mmol l^{-1}in *C. carnifex*, thereby maximized the Na diffusion gradient to the water (Dela-Cruz *et al.* manuscript in preparation). Maximum rates of Na loss were determined in crabs transferred to, and submerged in, containers with 1.4 l of water with the same salt composition as 30% seawater except for being free of Na and most of the Cl. Four groups of crabs (n=6 for each for each species) were assessed; with or without trapped lung air for the two species. The amount of Na appearing in the water was monitored at 5 min intervals for 40 min, and the increase in water Na concentration was determined. Haemolymph was sampled from crabs approximately 2 h prior to experiments so that the apparent permeability constant *K'* (Shaw, 1961; Sutcliffe, 1975; Greenaway, 1981) could be derived for Na. Sampling of the haemolymph prior to the experiment does not alter Na loss rates (Wood and Boutilier, 1985).

Unidirectional Na uptake was determined in *C. hirtipes* and *C. carnifex* previously held for 3 weeks with CIW for drinking and immersion, thus stimulating Na uptake mechanisms (Greenaway, 1989). Crabs were fitted with catheters as described above, and were submerged with or without lung air (n=6 for each species) in 1 l of water with the salt composition of CIW but containing Na at 15 mmol l^{-1}. The water was labeled with sufficient ^{22}Na (Amersham, Australia) as to provide 12,000 cpm for a 200 µl sample measured using a Hewlett-Packard scintillation counter. After transferring the towel-dried crabs into the experimental chambers, the water in the containers was sampled at 5 min intervals. Since the crabs invariably added small amounts of Na to the water, the exact start and end values of Na concentration were determined. This facilitated the calculation of unidirectional Na influx, J_{in} (µmol g^{-1} h^{-1}), using the special solution equation of Shaw (1963); $y=y_0\exp[(-Jt)/A]$ where y is the concentration of ^{22}Na in the water at time t, y_0 is the concentration of ^{22}Na at time zero and A is the amount of Na in the water. The total Na concentration of haemolymph and water samples was measured by atomic absorption spectrophotometry (AAS; GBC 906, GBC Melbourne). To suppress interference, samples were diluted with 5.9 mmol l^{-1} $CsCl_2$.

Statistical analysis

Data were analysed using SYSTAT 6.0 by ANOVA for fully independent designs and by ANOVAR for repeated measures experiments. Homogeneity of variances was confirmed using Bartlet's χ^2 test. Where ANOVA showed significant differences ($P<0.05$) from control groups, post-hoc testing was by Tukey's HSD or Contrast analysis.

Results

Lung air and O_2 uptake

The total lung volume was 18.8±3.4 and 15.5±1.77 ml in *C. hirtipes* and *C. carnifex* respectively and of this only 7–10% was released in the initial 5 min (Fig. 1). In *C. hirtipes* a further 5% was released in the following 25 min, but in both species 85% or more of the air was retained as a trapped bubble while submerged.

There were large changes in composition of the lung air during submersion of both species (Fig. 2). In air-breathing *C. carnifex* the lung air was 20.9% depleted of O_2 with respect to atmospheric (PO_2=20.7 kPa; 156 torr), while in *C. hirtipes* the reduction was 11.3%. After 30 min of submergence the PO_2 was markedly reduced to 3.52±0.30 kPa in *C. carnifex* and to 2.79±0.78 in *C. hirtipes*; reductions of 12.88 kPa and 15.68 kPa, or 79% and 85% of the available O_2 stored in the air bubbles, respectively. In addition to extreme hypoxia, submersion promoted a hypercapnia in the lung air (Fig. 2). Interestingly, the PCO_2 of the lung air of air-breathing *C. hirtipes* (0.79±0.09 kPa) and *C. carnifex* (0.83±0.10 kPa) was relatively hypercapnic but after 30 min of submersion, increased to 1.75±0.19 and 1.03±0.14 kPa respectively (Fig. 2).

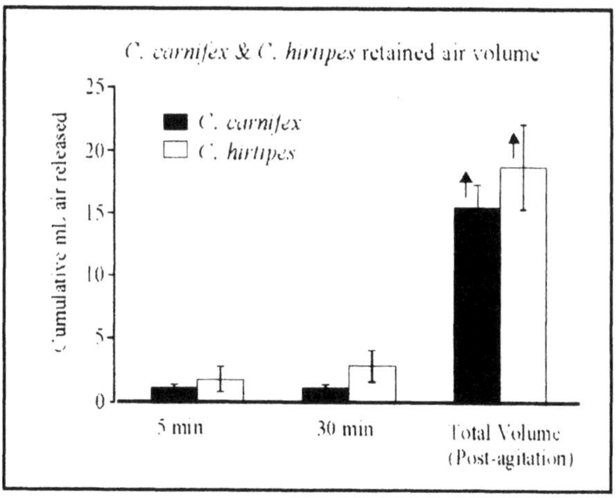

Fig. 1. The cumulative volume of air released from the lung/branchial chambers of *Cardisoma carnifex* and *C. hirtipes* after 5 and 30 min of submergence. The *final column* is the total volume of the lung lumen determined by agitating and inverting the crabs to release air trapped for the full 30 min of submergence. The *arrows* indicate a significant increase in volume, *n*=6 for each species

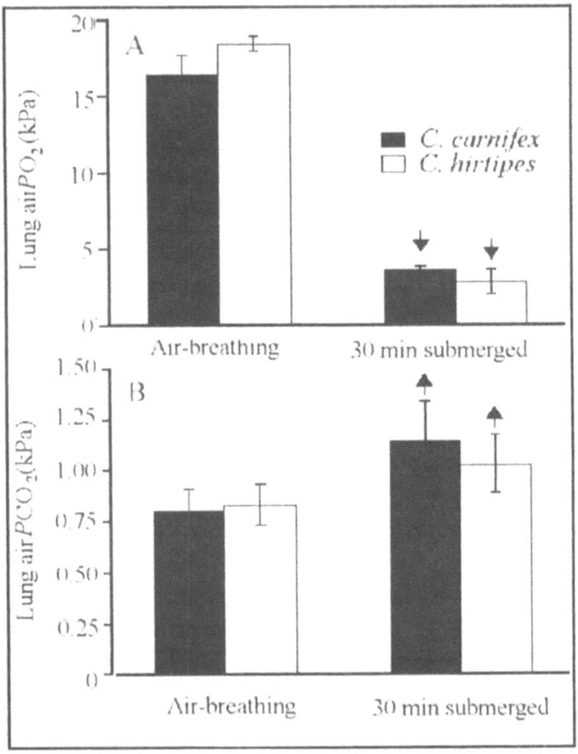

Fig. 2. A The partial pressure of O_2 (PO_2), and B CO_2 (PCO_2) of the air retained within the lumen of the lungs in *C. carnifex* and *C. hirtipes* either while breathing air or after 30 min of submergence. The *vertical arrows* indicate a significant change in partial pressure after submergence, n=6 for each species in each treatment

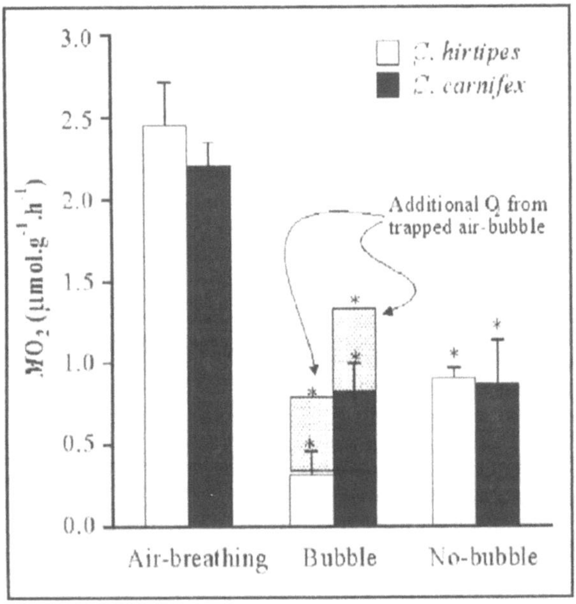

Fig. 3. The mass-specific O_2 uptake rate ($\dot{M}O_2$) of *C. hirtipes* and *C. carnifex* while breathing air or submerged, with or without air retained within the lung. The *open bars* show the contribution made by O_2 withdrawn from the trapped air during 30 min of submergence, converted to a mass specific rate (i.e. the molar amount withdrawn with respect to time and mass). n=6 for each species in each treatment. *A reduction compared to air-breathing animals. [Note: the data for air-breathing *C. hirtipes* were previously reported by Dela-Cruz and Morris (1997a)]

The MO_2 of air-breathing crabs was similar in the two *Cardisoma* at 2.45±0.27 μmol g^{-1} h^{-1} in *C. hirtipes* (previously reported in Dela-Cruz and Morris, 1997a) and 2.20±0.15 μmol g^{-1} h^{-1} in *C. carnifex* (Fig. 3). The O_2 uptake of both *C. hirtipes* and *C. carnifex* submerged with lung air decreased to 40% or less of the air-breathing rate (Fig. 3). Removing the air bubble such that the lung lumen was filled with water and the gills completely bathed did not improve MO_2. Taking the lung air volume into account together with the decrease in PO_2 of that air during submersion, the molar amount of O_2 taken from the bubble, and thus the contribution of stored lung O_2 during submersion, could be derived (Fig. 3). While this doubled the available O_2 used in *C. hirtipes* this was still equivalent to only a third of the air-breathing uptake. In *C. carnifex* using the O_2 in the bubble restored the effective MO_2 slightly more but only to 58% of the air-breathing rate and was not significantly improved over that of animals without trapped air bubbles (Fig. 3).

Haemolymph respiratory gas and metabolites

Submerging the crabs with lung air had significant effects on haemolymph respiratory gas status in both species and these were almost entirely unchanged by the removal of the air from the lung lumen (Table 1). The overall haemolymph CO_2 values were effectively the same in *C. hirtipes* with *C. carnifex*. The pulmonary and arterial haemolymph O_2 content were similar to each other within each species while air breathing. In both *C. hirtipes* and *C. carnifex* approximately 60% of the O_2 was removed during circulation of the haemolymph, 0.35–0.4 mmol l^{-1} comparing arterial (C_aO_2) and venous (C_vO_2) O_2 contents. The pulmonary haemolymph O_2 content (C_pO_2) and C_aO_2 of crabs submerged with lung air declined after 30 min and were no longer significantly elevated above C_vO_2 which was 0.24 mmol l^{-1} in *C. carnifex* and 0.38 mmol l^{-1} in *C. hirtipes* and, in both cases, unchanged from the C_vO_2 of air-breathing crabs (Table 1). Submerging crabs with the lung air removed did not assist in maintaining either C_pO_2 or C_aO_2 and in *C. hirtipes* may have lowered CO_2 further, since both C_aO_2 and C_vO_2 appeared to be 0.1 mmol l^{-1} lower in animals without trapped air in the lung lumen.

The internal hypoxia apparent in submerged crabs was even more obvious in the PO_2 values. Interestingly, the P_aO_2 in the haemolymph of air-breathing *C. hirtipes* (10.2 kPa) was twice that of *C. carnifex* (5.3 kPa) although the P_vO_2 values were similar (0.89 and 0.77 kPa respectively; Table 1). Submerging the crabs markedly reduced P_aO_2 to 2.1 kPa or less, which removed any significant arterial-venous PO_2 difference. As evidenced by the O_2 content values, removing the air bubble at the onset of emergence did not facilitate any improvement in haemolymph oxygenation.

Submergence of *Cardisoma* reduced the haemolymph CO_2 (Table 1), despite the developing hypercapnia in the air of the lung lumen, especially in *C. carnifex* in which P_aCO_2 declined from 1.95 to 1.08 kPa. Consequently, sub-

Table 1. Haemolymph respiratory gas status and pH in *C. hirtipes* and *C. carnifex* either breathing air or after 30 min submerged with trapped lung air, or after 30 min of submergence with all lung air completely removed. Oxygen and carbon dioxide contents (in mmol l^{-1}), and pH, are provided for efferent pulmonary haemolymph as well as for arterial and venous samples. The partial pressures are as kPa for arterial and venous samples (n=6–8)

	C_pO_2	C_aO_2	C_vO_2	C_pCO_2	C_aCO_2	C_vCO_2	pH_p	pH_a	pH_v	P_aO_2	P_vO_2	P_aCO_2	P_vCO_2
C. carnifex	(mmol L^{-1})									(kPa)			
Air-breathing	0.67 ± 0.17	0.82 ± 0.18	0.37 ± 0.05*+	18.58 ± 1.46	18.87 ± 1.11	19.07 ± 1.32	7.591 ± 0.030	7.569 ± 0.047	7.546 ± 0.037	***5.28*** ± 1.32	0.77 ± 0.23+	1.94 ± 0.23	2.05 ± 0.20
Submerged with lung air	0.21 ± 0.05#	0.39 ± 0.05#	0.24 ± 0.06	8.53 ± 1.23#	11.07 ± 1.16#	8.88 ± 0.75	7.594 ± 0.037	7.591 ± 0.027	7.569 ± 0.041	1.37 ± 0.15#	1.68 ± 0.32	1.49 ± 0.39	***0.70*** ± 0.13#
Submerged without lung air	0.33 ± 0.05#	0.30 ± 0.05#	0.25 ± 0.08	10.88 ± 1.01#	10.97 ± 1.40#	11.53 ± 1.34#	7.581 ± 0.019	7.575 ± 0.030	7.496 ± 0.029	1.32 ± 0.27#	1.26 ± 0.21	1.80 ± 0.24	1.97 ± 0.29
C. hirtipes													
Air-breathing	0.92 ± 0.09	0.86 ± 0.13	0.36 ± 0.08*+	15.20 ± 1.8	16.30 ± 1.57	16.70 ± 0.9	7.595 ± 0.034	7.622 ± 0.019	7.599 ± 0.028	***10.23*** ± 2.67	0.89 ± 0.05+	1.95 ± 0.17	2.13 ± 0.17
Submerged with lung air	0.31 ± 0.06#	0.54 ± 0.08#	0.38 ± 0.06	11.40 ± 1.31#	10.77 ± 1.67#	11.67 ± 1.65	7.628 ± 0.032	7.566 ± 0.031	7.468 ± 0.031	2.06 ± 0.39#	1.46 ± 0.47	1.08 ± 0.25#	***1.49*** ± 0.28#
Submerged without lung air	0.30± ± 0.05#	0.44 ± 0.02#	0.27 ± 0.02	8.80 ± 0.37#	8.77 ± 0.63#	8.93 ± 1.03	7.652 ± 0.026	7.625 ± 0.024	7.549 ± 0.037	1.64 ± 0.12#	1.84 ± 0.16	0.90 ± 0.16#	1.24 ± 0.07#

merged crabs had CO_2 values that were only 46–59% those of air-breathing animals in *C. carnifex* and 66–75% in *C. hirtipes* (Table 1). There was no evidence that submergence with the lung lumen (branchial chamber) completely flooded elicited any significant improvement of CO_2 excretion since P_aCO_2 remained at 1.80 and 0.90 kPa in *C. carnifex* and *C. hirtipes* respectively (Table 1). There was an apparent further decline in CCO_2 in *C. hirtipes* with the trapped air removed, to 53–58% of that in air-breathing crabs but this seemed independent of PCO_2.

Despite the relatively large changes in CO_2 associated with submergence, with or without a trapped air bubble, there were no significant changes in haemolymph pH in either species of *Cardisoma* (Table 1). Neither were there differences in the pH of haemolymph comparing pulmonary, arterial and venous samples for either species under each of the three different conditions.

The concentration of l-lactate in the haemolymph remained low (1.12–1.32 mmol l^{-1}) and was essentially the same in the two *Cardisoma* when submerged with a trapped air bubble in the lung (Table 2). However, l-lactate increased to 3.5 mmol l^{-1} in *C. hirtipes*, but not *C. carnifex*, when submerged without air within the lung. The haemolymph of *C. hirtipes* contained more urate but less glucose than that of *C. carnifex* (Table 2). In addition, submerging *C. carnifex* without trapped air in the lung induced a marked increase in haemolymph glucose compared to crabs submerged for 30 min with air trapped within the lung.

The influence of lung air on Na flux in submerged crabs

In submerged *C. hirtipes* and *C. carnifex* with a retained air bubble the Na loss rate to Na-free water was only 2.6±0.55 and 2.3±0.34 μmol g^{-1} h^{-1} but when the air bubble was removed the rate increased to 3.9±0.42 and 5.0±0.48 μmol g^{-1} h^{-1} respectively (Fig. 4). While the loss rate was not significantly greater in *C. carnifex* than in *C. hirtipes* the magnitude of the increase was larger (Fig. 4). The apparent permeability constant *K'* can be calculated as [ion loss rate (in mmol l^{-1} haemolymph) per hour divided by the haemolymph concentration (in mmol l^{-1})]; the extracellular/haemolymph space was taken as 33.2% for *C. hirtipes* and 24.7% for *C. carnifex* (Dela-Cruz and Morris, unpublished). The value of *K'* and thus apparent permeability was approximately doubled by bubble removal in *C. carnifex* from 0.032±0.005 to 0.067±0.009 but in *C. hirtipes* *K'* was 0.032±0.012 in animals with trapped air and similarly 0.039±0.004 after 30 min of lung (branchial chamber) flooding.

Unidirectional Na uptake (J_{in}) was determined in *Cardisoma* acclimated to freshwater (Fig. 5). The Na uptake rate of *C. hirtipes* submerged with lung air

[#] Different from air-breathing crab of same species
Values in ***bold italics*** indicate differences between C. carnifex and C. hirtipes
[*] Different from pulmonary value
[+] Different from arterial value

was only 0.90±0.30 μmol g^{-1} h^{-1} compared to the 1.74±0.54 μmol g^{-1} h^{-1} crabs submerged without the trapped air. Similarly, *C. carnifex* had a J_{in} of 1.88±0.61 μmol g^{-1} h^{-1} with air in the lung but 2.79±0.88 μmol g^{-1} h^{-1} with the air bubble removed (Fig. 5). However, there were considerable differences in the rates of individuals which masked both the difference between species ($F_{1,19}$=3.56, P=0.07) and the effect of bubble removal ($F_{1,19}$=2.62, P=0.12).

Discussion

Respiratory and acid–base consequences of submergence and trapped lung air

C. hirtipes and *C. carnifex* retain most of the air within the lung lumen during submersion, confirming previous field (Adamczewska and Morris, 1996; Morris and Adamczewska, 1996) and laboratory observations (Dela-Cruz and Morris, 1997a). The PO_2 and PCO_2 in the lung air of *C. hirtipes* and *C. carnifex* were indistinguishable from the PO_2 of 18.7 kPa and PCO_2 of 0.66 kPa previously obtained for „branchial chamber air" in *C. carnifex* (Wood and Randall, 1981). The large hypoxic hypercapnia that develops within the lung lumen signifies there was little gas exchange between the air space and the water, and minimal ventilation of the chamber. Both species showed greatly reduced but nonetheless appreciable O_2 uptake from the water, possibly facilitated by the lowered PO_2 within the lung air space. The MO_2 of immersed *C. hirtipes* (2.45 μmol g^{-1} h^{-1}; see also Dela-Cruz and Morris, 1997a) and *C. carnifex*

Table 2. Haemolymph concentration of L-lactate, urate and glucose in C. hirtipes and C. carnifex after 30 min of submergence, either with air within the lung lumen or with that air removed (n=6 for each species)

	L-lactate (mmol l^{-1})	Urate (mmol l^{-1})	Glucose (mmol l^{-1})
C. hirtipes			
Submerged with lung air	1.12 ± 0.15	***0.116*** ± 0.153	***0.85*** ± 0.19
Submerged without lung air	3.46 ± 0.81#	***0.073*** ± 0.073	***1.05*** ± 0.22
C. carnifex			
Submerged with lung air	1.32 ± 0.37	***0.028*** ± 0.065	***1.19*** ± 0.35
Submerged without lung air	0.87 ± 0.41	***0.030*** ± 0.083	***1.83*** ± 0.31#

Effect of removing lung air
Values in ***bold italics*** indicate a difference between corresponding values in the two species

(2.20 μmol $g^{-1} h^{-1}$) were also similar to previous values for these species (Wood and Randall, 1981) and to values for other obligate air-breathing crabs (Adamczewska and Morris, 1994 for review figure). Submersion reduced the MO_2 of *C. carnifex* by 60% and similarly by nearly 80% in *C. hirtipes*, as previously reported for the latter species (Dela-Cruz and Morris, 1997a). Flooding the lung failed to improve MO_2, perhaps because of minimal ventilation during submergence. This may not be true for all *Cardisoma* since O'Mahoney and Full (1984) found that *C. guanhumi*, in which the lung air was removed, maintained MO_2 in air and water at equal rates but only at the expense of a 300% increase in ventilation. Conversely, clearly obligate land crabs, such as *Gecarcinus lateralis*, reduced MO_2 in water by almost 50% compared to in air (Taylor and Davies, 1981, 1982; O'Mahoney and Full, 1984). Thus, both *C. hirtipes* and *C. carnifex* appear more committed to breathing air than *C. guanhumi*. Expelling the air would not improve the respiratory status of the two species and in fact the lung air store added appreciably to the O_2 available during the 30 min submergence.

The maximum P_aO_2 in air-breathing *C. hirtipes* was high but consistent with previous field determinations (Adamczewska and Morris, 1996) while greater than laboratory values (Farrelly and Greenaway, 1994; Dela-Cruz and Morris, 1997a). The arterial PO_2 in air-breathing *C. carnifex* was 50% of that in *C. hirtipes* suggesting relatively impaired diffusive uptake of O_2. However, in both *C. hirtipes* and *C. carnifex*, the pulmonary and arterial haemolymph of air-breathing crabs was apparently 100% saturated with O_2 (Burggren and McMahon, 1981; Farrelly and Greenaway, 1994; Adamczewska and Morris, 1996; Morris and Adamczewska, 1996; Dela-Cruz and Morris, 1997a). The similarity in pulmonary and arterial values suggests that either the gills were little perfused or, perhaps, that the gills and lungs were functioning similarly in O_2 uptake from air.

Field studies of *C. carnifex* (Morris and Adamczewska, 1996), and both field and laboratory studies of *C. hirtipes* (Adamczewska and Morris, 1996; Dela-

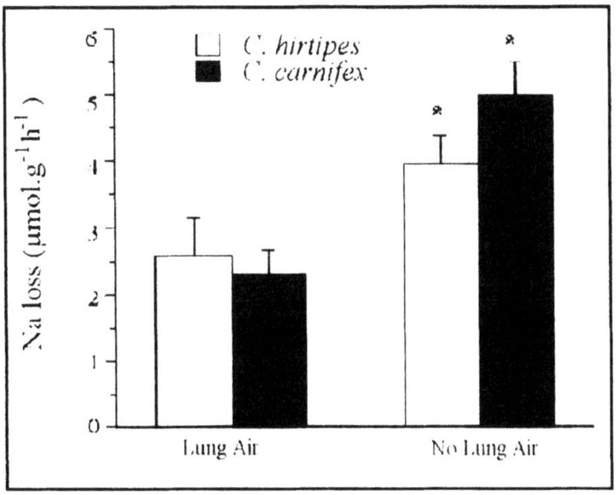

Fig. 4. The maximum Na loss rate, as μmol $g^{-1} h^{-1}$, from *C. hirtipes* and *C. carnifex* submerged with or without an air bubble retained within the lung. *An increase in loss associated with removal the retained air bubble. Both *Cardisoma* species were acclimated to drinking water of 30% seawater to maximize loss rates, n=6 for each species in each treatment

Cruz and Morris, 1997a) showed that 30 min of submergence rapidly depletes haemolymph O_2. In the current study, the pulmonary and arterial CO_2 both declined to near the C_vO_2 after 30 min under water, which contrasts with *Gecarcinus lateralis* in which both arterial and venous haemolymph O_2 content decreased to similar extent (Taylor and Davies, 1982). The reduction in lung air PO_2 to 3 kPa or less meant that the maintenance of P_aO_2 was impossible. The gradient for inward diffusion of O_2 from the lung lumen into the haemolymph declined from 8.27 to 0.79 kPa in *C. hirtipes* and from 11.22 to 2.15 kPa in *C. carnifex*. Consequently, P_vO_2 must decline to sustain any inward diffusion of O_2 but eventually failed in *Cardisoma* since the arterial–venous difference declined to zero. The haemocyanin-bound O_2 (Hc-O_2) has been previously calculated to provide O_2 for approximately 5 min in a 400 g *C. hirtipes* (Adamczewska and Morris, 1996) and 8 min in 300 g *C. carnifex* (Morris and Adamczewska, 1996). These stores in crabs submerged for 30 min with trapped air bubbles (Fig. 3) were depleted at a rate of 0.23 and 0.31 μmol $g^{-1} h^{-1}$ *C. hirtipes* and *C. carnifex*, respectively, effectively supplementing the MO_2 which was still well below the air-breathing MO_2, of *C. hirtipes* especially. In the absence of any appreciable anaerobiosis in *C. hirtipes*, Adamczewska and Morris (1996) concluded that the crabs become relatively hypometabolic; a very different strategy to the threefold to fourfold elevation in ventilatory work suggested by O'Mahoney and Full (1984) for submerged *C. guanhumi*.

Submerging *C. hirtipes* and *C. carnifex* without trapped air, such that the water bathed the entire branchial structure, failed entirely to relieve the submergence hypoxia. This is quite at odds with both Cameron (1981) and O'Mahoney and Full (1984), who suggested that this allowed respiratory exchange with water. Nonetheless, neither *C. hirtipes* or *C. carnifex* showed any improvement in either the haemolymph PO_2 or consequently the O_2 content. Conversely, the CO_2 data for *C. hirtipes* support the suggestion that removing the air bubble actually exacerbates the hypoxia, since haemolymph lactate concentrations increased to 3.5 mmol l^{-1}. The effect of submergence on haemolymph metabolites has been shown to be variable between species and depend on circumstance. For example, in *C. hirtipes* neither urate nor the l-lactate concentration increased in crabs submerged in the field (Adamczewska and Morris, 1996) whereas urate, but not L-lactate, did increase during submergence in the laboratory (Dela-Cruz and Morris, 1997a). Similarly, in the current study lactate increased in *C. hirtipes* only when the air bubble was removed from the lung lumen. Field determinations of *C. carnifex* also showed urate to be unaltered in the haemolymph during submersion, but, in contrast to *C. hirtipes*, there were some increases in haemolymph l-lactate levels although urate was unchanged by submersion (Morris and Adamczewska, 1996). In both field and laboratory determinations the submersion of *C. hirtipes* promoted a hypoglycaemia whereas *C. carnifex* showed evidence of a hyperglycaemia (Morris and Adamczewska, 1996) which was evident in the present study only once the air bubble was removed. Thus, in *C. hirtipes* re-

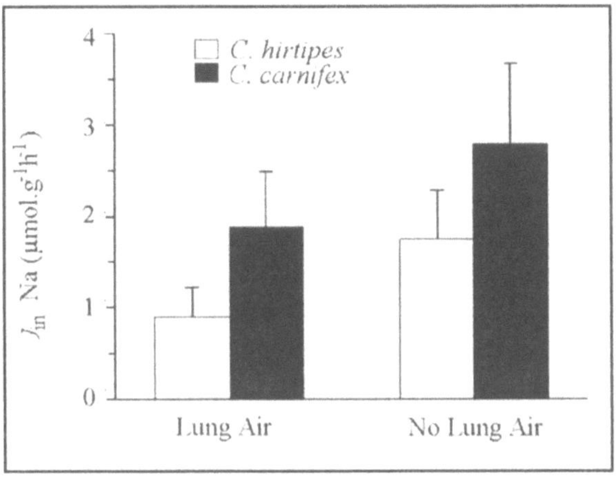

Fig. 5. The unidirectional Na uptake rate (J_{in}), as µmol g^{-1} h^{-1}, by *C. hirtipes* and *C. carnifex* submerged with or without an air bubble retained within the lung. ANOVA provided P=0.07 for the difference between species and P=0.12 for the effect of bubble removal. Both *Cardisoma* species were acclimated to drinking CIW to maximize uptake rates, n=6 for each species in each treatment

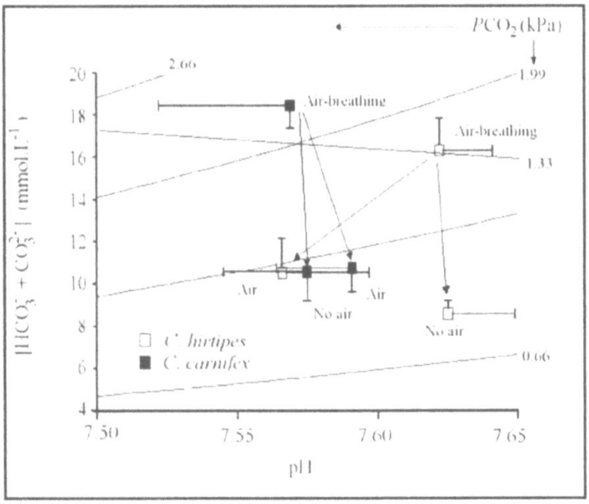

Fig. 6. The pH/HCO_3^- diagram for *C. hirtipes* and *C. carnifex* in each of the three treatments: air-breathing, submerged with air retained in the lung (bubble) and with the air removed (no bubble). The slope of the non-bicarbonate buffer line was from the data of Dela-Cruz and Morris (1997) and the PCO_2 isopleths derived from the Henderson-Hasselbalch relationship using a pK'=6.025 and an αCO_2=0.2267 µmol l^{-1} Pa^{-1}

moval of the lung air, and therewith an O_2 store that was otherwise used during submergence, made a crucial difference in the recruitment of anaerobiosis, which does not normally occur in *C. hirtipes*, although it was observed by Shah and Herreid (1978) for *C. guanhumi*. The lactacidosis in *C. carnifex* measured in crabs submerged in the field caused Morris and Adamczewska (1996) to conclude that this species differs from *C. hirtipes* in that it maintains its metabolic rate by supplementary anaerobiosis. However, there was no such anaerobiosis in the present laboratory study of *C. carnifex*. Given that 30 min of submergence has proven to be close to the aerobic limit for both species, and the extremely low O_2 content in the haemolymph, it seems reasonable that small differences, for example in temperature, may determine the duration of submergence that can be attained without initiating anaerobiosis.

Haemolymph acidosis can arise because of accumulation of respiratory CO_2 as well as through metabolic acidosis. The haemolymph PCO_2 and CCO_2 were generally not different comparing *C. hirtipes* and *C. carnifex*, and the PCO_2 values in air-breathing crabs were essentially the same as previously obtained values for *C. carnifex* (1.92 kPa, Randall and Wood, 1981; 1.90 kPa, Morris and Adamczewska, 1996) and for *C. hirtipes* (1.81 kPa, Farrelly and Greenaway, 1994; >1.60 kPa, Adamczewska and Morris, 1996). Submergence of *C. hirtipes* and *C. carnifex*, despite air trapped within the lung, induced a haemolymph hypocapnia in both species, suggesting some facilitation of CO_2 excretion by bathing water. Laboratory studies of *C. carnifex* showed a 2.5-fold increase in ventilation volume when in water (McMahon and Burggren, 1988 for review) and as a consequence reduced haemolymph PCO_2 and CCO_2, and thus a respiratory alkalosis (Cameron, 1981), in common with nearly every species able to inhabit air and water (McMahon and Burggren, 1988). However, this was not the case for either *C. hirtipes*, which exhibited a marked metabolic alkalosis (Adamczewska and Morris, 1996), or *C. carnifex* voluntarily submerged in the field, which showed a respiratory acidosis (Morris and Adamczewska, 1996). This provides further evidence that submerged *Cardisoma* ventilate very little and that gas exchange is primarily with a progressively hypoxic-hypercapnic air bubble. Excretion of CO_2 was therefore severely impaired in submerged *Cardisoma*, thus the gradient for outward diffusion into the lung lumen from the haemolymph declined from 1.11 to 0.46 kPa in *C. carnifex*, and in *C. hirtipes* changed from an outward 1.16 kPa to an inward gradient of 0.67 kPa, as a result of the hypercapnia in the lung air.

The acid-base status, the influence of submersion, the effect of removing the trapped air bubble from the lung and the differences between *C. carnifex* and *C. hirtipes* are best appreciated with the aid of a pH/HCO_3^- diagram (Fig. 6). Under normal circumstances *C. carnifex* exist in a state of relative respiratory acidosis compared with *C. hirtipes* (Fig. 6). This cannot be ascribed to higher CO_2 production rates in *C. carnifex* since the two species have similar respiration rates. Neither can it be due a relatively hypercapnic lung lumen; thus, *C. carnifex* must experience relatively greater diffusion resistance to CO_2 excretion, consistent with the apparent relative resistance to inward O_2 diffusion (above). Thus, *C. hirtipes* is better than *C. carnifex* at breathing air. The submergence of *C. carnifex* and *C. hirtipes* promoted some metabolic acidosis which in *C. carnifex* especially was compensated by a respiratory alkalosis (Fig. 6). The source of the metabolic acid is not obvious but the pH changes were slight. *C. hirtipes* submerged in the field experienced net ion loss (Greenaway, 1989; Adamczewska and Morris, 1996) and changes in strong ion concentrations within the haemolymph could account for the slight acidosis. The relative hypocapnia that occurred in both species after 30 min submerged indicates that some improved CO_2 excretion to the water occurs, even with retained air in the lung. Similarly, the compensatory respiratory alkalosis in *C. carnifex* is a consequence of improved CO_2 excretion. That this compensation is less apparent in submerged *C. hirtipes* indicates relatively less facilitation of CO_2 excretion and perhaps less contact of the exchange surfaces with water. Flooding the lung/branchial chamber of *C. carnifex* did little to further enhance the compensatory respiratory alkalosis, whereas in *C. hirtipes* the alkalosis was really apparent only once the lumen was flooded and a larger epithelial surface area was bathed (Fig. 6).

Thus, the air bubble retained by the two *Cardisoma* investigated does not hinder O_2 uptake by submerged crabs, since its removal does not alleviate the internal hypoxia, but instead it may provide an important O_2 reservoir. Under laboratory conditions expelling the air bubble may assist some CO_2 excretion, but apparently not under field circumstances, and to a greater extent in *C. carnifex* than in *C. hirtipes*.

Immersion avoidance and lung air retention as mechanisms to minimize ion loss

Under field conditions *C. hirtipes* lost Ca and Mg while submerged (Adamczewska and Morris, 1996) and in the laboratory lost salt to freshwater (Greenaway, 1989; Dela-Cruz and Morris, 1997a). Adamczewska and Morris (1996) suggested that the trapped lung air, by reducing contact between the water and permeable branchial and branchiostegal epithelia, could mitigate ion losses to freshwater and lower regulative costs in *C. hirtipes*, and thereby explain the propensity to avoid prolonged submersion. Similarly, *C. carnifex* given only fresh water lost K, Ca and Mg (Wood and Boutilier, 1985) but salt regulation is perhaps less problematic for *C. carnifex* since it can visit the ocean to replace salts lost to lower salinity water in the burrows (Morris and Adamczewska, 1996).

C. hirtipes and *C. carnifex* submerged in Na-free CIW in the laboratory lost Na 33% and 54% (respectively) more rapidly when the air trapped within the lung lumen (branchial chamber) was removed. There was thus a greater effect on Na loss rates in *C. carnifex*, possibly since the Na concentration gradient was 13% greater than for *C. hirtipes*. The rates were high compared to freshwater crabs, e.g. *Potamonautes warreni* at 0.22 $\mu mol\ g^{-1} h^{-1}$ (Morris and van Aardt, 1998) and 0.07–0.20 $\mu mol\ g^{-1}\ h^{-1}$ *Holthuisana transversa* (Greenaway, 1981), and represent significant Na losses which are greatly exacerbated by loss of the protecting air bubble, which must mean a marked increase in energetic requirements. However, the crabs used in the Na loss measurements were pre-acclimated to 30% SW to emphasize any effect of the air bubble on loss rates. The loss rates in freshwater-acclimated crabs are much lower, 0.12 and 0.52 $\mu mol\ g^{-1} h^{-1}$ for *C. hirtipes* and *C. carnifex* respectively, due to reduced permeability (Dela-Cruz, Morris and Greenaway, unpublished) and the actual consequential costs may be lower.

The salt regulation costs during immersion, e.g. compensatory Na uptake, can only be reduced by decreased permeability. Exact measurements of whole-animal permeability are often difficult to obtain since they depend not only on the diffusion resistance of the epithelia but also on their surface area, but estimates of *K'* can provided good estimates of apparent Na permeability (Mantel and Farmer, 1983). The value of *K'* increased when the air bubbles were removed, in *C. carnifex* by 107% and in *C. hirtipes* by 22%, implying that the greater *K'* of crabs submerged without trapped branchial air must be due to the increased exposure of the gills and/or branchiostegites to water.

The unidirectional Na uptake measurements on CIW-acclimated *Cardisoma* cannot be directly compared to the loss rates since acclimation salinity also alters J_{in} (Dela-Cruz *et al.* manuscript in preparation). However, the uptake rates were highest in animals in which the lung and gills were flooded. A higher rate of uptake in *C. carnifex* than in *C. hirtipes* is consistent with compensation needed to accommodate the increased Na loss rate and apparent permeability. It is difficult to determine to what extent these data indicate true compensation. It seems probable that in crabs with trapped lung air that part of, or even all, the gills would be in contact with water, hence the Na exchange in submerged crabs, and that expelling the air simply increases the surface area bathed to include the lung also, thereby increasing the surface involved both in diffusive ion loss and in compensatory pumping.

Neither species of *Cardisoma* investigated seems to be a truly amphibious crab since they both appear to eventually asphyxiate under water (Adamczewska and Morris, 1996; Morris and Adamczewska, 1996). *C. hirtipes* and *C. carnifex* naturally trap air within the lung lumen, rather than expose the lining and the gills to water. This trapped air acts as an O_2 store and a CO_2 repository. Expelling the trapped air fails to significantly improve any aspect of respiration and indeed removes a usable O_2 store. In many previous studies of *Cardisoma* the crabs were probably able to breathe air facultatively at the water surface. *Cardisoma* submerged in freshwater lose Na but when the air is removed from the lung lumen (branchial chamber) Na loss, permeability and Na uptake are increased. Retention of trapped air has the major effect of reducing Na turnover and associated pumping costs. Under normal humid conditions in tropical and sub-tropical habitats the only identifiable reason for *Cardisoma* to be so closely associated with water is for the excretion of nitrogen (Dela-Cruz and Morris, 1997b). These conclusions need to be extended and tested by investigation of other *Cardisoma* species and by further field determinations under natural conditions.

Acknowledgements. We would like to thank Agnieszka Adamczewska, Peter Greenaway and Stuart Linton for assistance in collecting animals. We also thank Professor Peter Greenaway for helpful discussions and Dr. Adamczewska for critical reading of the manuscript. Thanks go also to the Conservators and Staff of Parks Australia North, on Christmas Island and on Cocos Island for their hospitality and help. This work and a Post-Graduate Scholarship to J.D.-C. were supported by ARC grants to S.M.

References

Adamczewska, A.M., Morris, S. (1994) Exercise in the terrestrial Christmas Island red crab *Gecarcoidea natalis*. I. Blood gas transport. *J. Exp. Biol.* **188**:235–256

Adamczewska, A.M., Morris, S. (1996) The respiratory gas transport, acid-base state, ion and metabolite status of the Christmas Island Blue Crab, *Cardisoma hirtipes* (Dana) assessed *in situ* with respect to immersion. *Physiol. Zool.* **69**:67–92

Bridges, C.R., Bicudo, J.E.P.W., Lykkeboe, G. (1979) Oxygen content measurement in haemolymph containing haemocyanin. *Comp. Biochem. Physiol.* **62A**:399–409

Burggren, W.W., McMahon, B.R. (1981) Haemolymph oxygen transport, acid-base status and hydromineral regulation during dehydration in three terrestrial crabs, *Cardisoma*, *Birgus* and *Coenobita*. *J. Exp. Zool.* **218**:1–5

Burggren, W., Pinder, A., McMahon, B.R., Wheatly, M., Doyle, M. (1985) Ventilation, circulation and their interactions in the land crab, *Cardisoma guanhumi*. *J. Exp. Biol.* **117**:133–154

Cameron, J.N. (1981) Acid-base responses to changes in CO_2 in two Pacific crabs: *Birgus latro*, and the mangrove crab *Cardisoma carnifex*. *J. Exp. Zool.* **218**:65–73

De Leersnyder, M., Hoestlandt, H. (1963) Premiers donnes sur la regulation osmotique et la regulation ionique du crabe terreste *Cardisoma armatum* Herklots. *Cah. Biol. Mar.* **4**:211–218

Dela-Cruz, J., Morris, S. (1997a) Respiratory, acid-base, and metabolic responses of the Christmas Island Blue Crab, *Cardisoma hirtipes* (Dana) to simulated immersion. *Physiol. Zool.* **70**:100–115

Dela-Cruz, J., Morris, S. (1997b) Water and ion balance, and nitrogenous excretion as limitations to aerial excursion in the Christmas Island Blue Crab, *Cardisoma hirtipes* (Dana). *J. Exp. Zool.* **279**:537–548

Farrelly, C.A., Greenaway, P. (1992) Morphology and ultrastructure of the gills of terrestrial crabs (Crustacea, Gecarcinidae and Grapsidae): adaptations for air breathing. *Zoomorphology* **112**:39–49

Farrelly, C.A., Greenaway, P. (1993) Land crabs with smooth lungs: Grapsidae, Gecarcinidae and Sundathelphusidae; ultrastructure and vasculature. *J. Morphol.* **215**:1–16

Farrelly, C.A., Greenaway, P. (1994) Gas exchange through the lungs and gills in air-breathing crabs. *J. Exp. Biol.* **187**:113–130

Gifford, C.A. (1962) Some observations on the general biology of the land crab *Cardisoma guanhumi*. *Am. Zool.* **123**:207–223

Greenaway, P. (1981) Sodium regulation in the freshwater/land crab *Holthuisana transversa*. *J. Comp. Physiol. B* **142**:451–456

Greenaway, P. (1989) Sodium regulation and adaptation to freshwater in the amphibious crab *Cardisoma hirtipes*. *Physiol. Zool.* **62**:639–653

Hartnoll, R.G. (1988) Evolution, systematics, and geographical distribution. In: Burggren, W.W., McMahon, B.R. (eds) *Biology of the land crabs*.. Cambridge University Press, New York, pp. 6–53

Herreid, C.F., Gifford, C.A. (1963) The burrow habitat of the land crab *Cardisoma guanhumi* Latreille. *Ecology* **44**:773–775

Mantel, L.H., Farmer, L.L. (1983) Osmotic and ionic balance. In: Mantel L.H. (ed) *The Biology of Crustacea*, vol. 5. *Internal anatomy and physiological regulation*. Academic, New York, pp. 54–143

McMahon, B.R., Burggren, W.W. (1988) Respiration. In: Burggren, W.W., McMahon, B.R. (eds) *Biology of the land crabs*. Cambridge University Press, New York, pp. 249–297

Morris, S., Adamczewska, A.M. (1996) *In-situ* immersion of *Cardisoma carnifex*. *J. Exp. Mar. Biol. Ecol.* **206**:149–164

Morris, S., van Aardt, W.J. (1998) Salt and water relations, and nitrogen excretion in the amphibious freshwater crab *Potamonautes warreni* with respect to immersion. *J. Exp. Biol.* **201**:883–893

Morris, S., Curtin, A.L., Thompson, M.B. (1994) Heterothermy, torpor, respiratory gas exchange, water balance and the effect of feeding in Gould's long-eared bat *Nyctophilus gouldi* (Tomes 1858). *J. Exp. Biol.* **197**:309–335

O'Mahoney, P.M., Full, R.J. (1984) Respiration of crabs in air and water. *Comp. Biochem. Physiol.* **79**:275–282

Pinder, A.W., Smits, A.W. (1993) The burrow microhabitat of the land crab *Cardisoma guanhumi*: respiratory/ionic conditions and physiological responses of crabs to hypercapnia. *Physiol. Zool.* **66**:216–236

Randall, D.J., Wood, C.M. (1981) Carbon dioxide excretion in the land crab (*Cardisoma carnifex*). *J. Exp. Zool.* **218**:37–44

Shah, G.M., Herreid, C.E. (1978) Heart rate of the land crab, *Cardisoma guanhumi* (Latreille), during aquatic and aerial respiration. *Comp. Biochem. Physiol.* **60**:335–341

Shaw, J. (1961) Sodium balance in *Eriocheir sinensis* (M. Edw.). The adaptation of the Crustacea to freshwater. *J. Exp. Biol.* **38**:153–162

Shaw, J. (1963) Kinetic aspects of ion regulation in aquatic animals. In: Carthy, J.D., Duddington, C.L. (eds) *Viewpoints in biology*. Butterworth, London, pp. 2, 163–201

Sutcliffe, D.W. (1975) Sodium regulation and adaptation to fresh water in the isopod genus *Asellus*. *J. Exp. Biol.* **61**:719–736

Taylor, A.C., Davies, P.S. (1981) Respiration in the land crab, *Gecarcinus lateralis*. *J. Exp. Biol.* **93**:197–208

Taylor, A.C., Davies, P.S. (1982) Aquatic respiration in the land crabs, *Gacarcinus lateralis* (Fréminville). *Comp. Biochem. Physiol.* **72A**:683–688

Tucker, V.A. (1967) Method for oxygen content and dissociation curves on microliter blood samples. *J. Appl. Physiol.* **23**:410–414
Wolcott, D.L. (1991) Integration of cellular, organismal, and functional ecological aspects of salt and water balance. *Mem. Qld. Mus.*31:229–239
Wolcott, T.G. (1992) Water and solute balance in the transition to land. *Am. Zool.* **32**:370–381
Wood, C.M., Boutilier, R.G. (1985) Osmoregulation, ionic exchange, blood chemistry and nitrogenous waste excretion in the land crab *Cardisoma carnifex*: a field and laboratory study. *Biol. Bull.* **169**:267–290
Wood, C.M., Randall, D.J. (1981) Oxygen and carbon dioxide exchange during excercise in the land crab (*Cardisoma carnifex*). *J. Exp. Zool.* **218**:7–16

Calcium, polarity and osmoregulation in *Fucus* embryos: one messenger, multiple messages

C. Brownlee · N.F.H. Manison · R. Anning

Marine Biological Association, The Laboratory, Citadel Hill, Plymouth PL1 2PB, U.K.

Received: 14 May 1998 / Accepted: 20 October 1998 / Published: 10 November 1998

Abstract. This review addresses the role of cytosolic Ca^{2+} in the regulation of embryonic polarity, polarized growth and osmotic control of cell volume with particular emphasis on the marine alga *Fucus*. Evidence is presented that the control of cell elongation in plants and algae requires the co-ordinated control of cell turgor and vesicle exocytosis, all of which are regulated by cytosolic Ca^{2+}. In the *Fucus* embryo, polarized growth of the rhizoid cell requires sustained elevation of Ca^{2+} at the rhizoid apex whereas osmotic cell volume control is effected by transient elevations of cytosolic Ca^{2+}. The mechanisms by which these signals are generated and their different downstream responses are discussed. Answers to questions of how specificity of signalling can be achieved by signal-response pathways which share common components will require more detailed knowledge of the interactions between different components and the spatial and temporal patterns of their activation.

Key words. Calcium · Cell volume control · *Fucus* · Growth · Polarity Signalling

Introduction

Plant development: interactions between pattern, morphogenesis and physiology

Plant shape and form during development can be considered to result from the acquisition and expression of pattern coupled to morphogenetic responses to external signals. Pattern in this context is defined as the spatial organization of tissues and organs and requires the interpretation of intrinsic or external polarity signals and differentiation of cells in response to positional cues (Van den Berg et al., 1995; Laux and Jurgens, 1997; Bouget and Brownlee, 1998). Morphogenetic responses are superimposed on overall pattern and can be de-

Correspondence to: C. Brownlee (e-mail: CBR@wpo.nerc.ac.uk)

fined as variations in the relative size, shape, degree of differentiation and growth direction of organs, tissues and cells in response to external signals. Thus morphogenesis results in environmentally determined changes in the shape and size of organs while basic body pattern (i.e. the positions of organs and tissues) generally remains unchanged. A good example is floral induction. In many species, the development of floral primordia occurs in response to external signals. The positions of the floral primordia in relation to other organs, however, are almost invariably fixed. Examples of morphogenetic signals include light quality, direction and photoperiod, gravity and mechanical stimuli. Plants are able to integrate a range of external signals and translate them into appropriate morphogenetic and physiological responses at the single-cell and whole-plant level. This requires the operation of sophisticated signalling machinery which clearly involves communication between a wide range of receptor, transduction and response elements. A good example of this is the control of stomatal aperture which is responsive to light, hormones (e.g. auxin, abscisic acid) and carbon dioxide (McAinsh et al., 1997). These stimuli are sensed by different receptor elements while the signal transduction pathways converge at the response level to regulate the flux of ions and water to and from the stomatal guard cells and, via control of turgor pressure and cell shape, effect control of stomatal aperture.

The following discussion addresses some of the problems which arise when a single second messenger is clearly a component of more than one signal transduction pathway resulting in different end responses. This problem is particularly relevant to the role of cytosolic Ca^{2+} as a second messenger. Ca^{2+} is involved in signalling in a wide range of stimulus-response pathways and understanding how specificity can arise from a single messenger is a major goal of plant signal transduction studies (McAinsh and Hetherington, 1998).

A variety of different systems are considered here, with examples of some recent work on the *Fucus* zygote which presents a system in which signalling involving cytosolic Ca^{2+} in relation to polarity, embryonic pattern and physiological responses can be studied in the same cell.

Regulation of cell growth

Plant cell growth requires, but is not regulated by, turgor

Understanding the processes which underlie plant cell growth is fundamental to understanding morphogenesis. Plant and algal cells grow by increasing the surface area of the plasma membrane and cell wall. Increases in cell size can occur in a variety of patterns. Elongation may occur over the whole surface of the cell or may be restricted to specific zones. An extreme example of this is polarized tip growth such as occurs in pollen tubes, root hairs, fungal hyphae and algal rhizoid cells. New cell wall and membrane material is continually incorporated at the growth point. Elongation in most plant cells requires internal positive hydrostatic pressure (turgor pressure) which is normally brought about by uptake of water into the vacuole and is thought in most cases to provide the driving force for elongation growth. However, the relationship between turgor and cell growth is not straightforward and varies between species. In maize coleoptiles, a curvilinear relationship exists between growth and turgor (Hohl and Schopfer, 1992) with a greater relative growth at higher turgor pressures. However, cell elongation rates can vary despite a constant turgor (Tomos and Pritchard, 1994) and it is likely that the main factor regulating the rate of cell expansion is the elasticity of the cell wall (Pritchard, 1994; Kutschera, 1996). It has been proposed that wall relaxation and subsequent expansion is one of the major regulators of cell turgor (Cosgrove, 1993; Kutschera and Kohler, 1994). Turgor clamp experiments with the brackish water charophyte alga *Chara* (Zhu and Boyer, 1992) showed that a threshold level of turgor was required for growth to occur. Above this threshold, however, growth was independent of turgor and relied more on metabolic energy, possibly reflecting the reliance of growth on the synthesis of wall polymers. In tip-growing cells, such as the lily pollen tube, turgor was shown to be regulated within narrow limits despite variations in cell elongation rate (Benkert et al., 1997). An extreme example of the independence of growth from turgor comes from studies of the water mould *Achlya*. In this species, growth has been shown to occur even in the absence of cell turgor (Money and Harold, 1992, 1993). It has been proposed that at low or zero turgor, apical growth in *Achlya* can be maintained via actin polymerization in much the same way as in animal cells (Harold et al., 1995). It follows from the above that morphogenetic changes in cell growth and elongation in response to external stimuli are likely to involve regulation of mechanisms other than those which regulate cell turgor.

Polarity and growth require regulation of targeted exocytosis

Exocytosis is essential for plant cell growth, providing the new wall material and plasma membrane necessary for the expanding cell. The growing pollen tube is a classic example of large-scale exocytosis directed to the growing apex (Steer and Steer, 1989). Inhibition of Golgi vesicle dynamics leads to inhibition of elongation. The absolute requirement of pollen tube growth on vesicle dynamics was shown in pollen tubes of *Nicotiana tobaccum* and *Petunia hybrida* where elongation was inhibited by the inhibitor of Golgi secretion brefeldin A (BFA) (Geitman et al., 1996). Pollen tube elongation is characterized by pulsatory variations in elongation rate and it has been proposed that such oscillations in growth rate may be a general property of all tip-growing cells (Holdaway-Clarke et al., 1997). While longer term elongation in *Nicotiana* and *Petunia* (Geitman et al., 1996) was inhibited by BFA, shorter pulse-like variations in the elongation rate which last 10–20 s were insensitive BFA, though they were inhibited by inhibitors of actin polymerization which probably reflects a requirement for the cytoskeleton in docking and/or fusion of vesicles already located at the tube apex. Zygotes of the marine alga *Fucus* are initially apolar following germination and first form then fix an axis of polarity in response to unidirectional light (Kropf, 1997). The polar axis defines the orientation of the first asymmetrical division which produces two cell types (rhizoid and thallus) with different fates. Fixation of the polar axis can thus be regarded as the initial step in determining cell fate. Axis fixation is known to require the presence of a cell wall (Kropf et al., 1988) and intact actin cytoskeleton (Brawley and Robinson, 1985). A role for targeted secretion in polar axis fixation has recently been invoked from experiments using BFA to block secretion (Shaw and Quatrano, 1996). During polarization, polar secretion of sulphated polysaccharide (fucoidin) occurs at the rhizoid pole. Treatment with appropriate concentrations of BFA during the period of axis fixation reversibly prevented polar secretion of sulphated fucoidin into the cell wall, axis fixation and rhizoid growth but did not prevent accumulation at the rhizoid pole of actin which also accompanies axis fixation. This suggests that localization of actin at the rhizoid pole alone is not sufficient to account for polar axis fixation and that asymmetries in the cell wall serve as positional markers for polarization. This is in agreement with the observation that cell wall factors are required for determination of rhizoid cell fate in two-celled embryos (Berger et al., 1994) and supports the hypothesis that these factors are inserted into the cell wall during establishment of the polar axis. Thus, in addition to the role in maintaining cell elongation, exocytosis of specific cell wall components appears to be an early step in establishing polarity and determining cell fate. This raises important questions concerning the nature of these secreted cell wall components. While these are likely to vary between different cell types, a good candidate for such factors in *Fucus* zygotes may be the asymmetrically distributed sulphated polysaccharide fucoidin. Recent evidence from carrot somatic embryos suggests the involvement of an arabinogalactoprotein acting specifically in the walls of cells destined to become somatic embryos (McCabe et al., 1997).

Multifunctional Ca^{2+} signals

Polarity and growth

In eukaryotic cells, resting levels of cytosolic Ca^{2+} are regulated within a very narrow concentration range around 0.1 µM. A feature which appears to be common in apically growing plant cells is the establishment of a region of sustained elevated cytosolic Ca^{2+}, reaching concentrations in the micromolar range or higher close to the growing apex. This has been verified in cells as diverse as pollen tubes (Rathore et al., 1991; Miller et al., 1992; Franklin-Tong et al., 1997), root hairs (Hermann and Felle, 1995; Felle and Hepler, 1997; Wymer et al., 1997), polarizing *Fucus* zygotes (Berger and Brownlee, 1993) and growing

Fucus rhizoids (Brownlee and Wood, 1986; Brownlee and Pulsford, 1988; Taylor et al., 1996). The available evidence strongly suggests that this apical Ca^{2+} gradient is essential for polarized growth. Thus, injection of pollen tubes with BAPTA-type buffers abolished the Ca^{2+} gradient and inhibited apical growth (Miller et al., 1992; Pierson et al., 1994). Ca^{2+} buffers have also been shown to prevent polarization in *Fucus* zygotes (Speksnijder et al., 1989) and to abolish the apical Ca^{2+} gradient (Taylor et al., 1996). In pollen tubes the apical Ca^{2+} gradient oscillates in phase with oscillations in growth (Messerli and Robinson, 1997). Peak pulsatile growth rates were shown to correlate with peaks in cytosolic Ca^{2+} at the tube apex. Ca^{2+} and growth oscillations were, however, shown to be out of phase with measurements of Ca^{2+} influx made with an extracellular vibrating Ca^{2+} electrode (Holdaway-Clarke et al., 1997). Ca^{2+} influx lagged behind Ca^{2+} oscillations by several seconds. Two possible models have been proposed to account for this, invoking either recharging of intracellular Ca^{2+} stores or external binding sites for Ca^{2+} in the cell wall (Holdaway-Clarke et al., 1997). Cytosolic Ca^{2+} has also been proposed to signal growth direction in *Agapanthus* pollen tubes. Treatments which caused asymmetrical elevations of Ca^{2+} in the pollen tube tip caused the tube to change growth direction to the side of the tube at which Ca^{2+} was highest (Malho and Trewavas, 1996).

Evidence of this kind implicates Ca^{2+} as having a major role in the regulation of exocytosis in tip-growing cells. Evidence is also accumulating for a more general role for Ca^{2+} in regulation of exocytosis in plants. Direct measurement of exocytosis, using patch-clamp measurements of capacitance in barley aleurone protoplasts (Zorec and Tester, 1992), maize coleoptiles (Thiel et al., 1994) and more recently in maize root cap cells (Carroll et al., 1998), indicate a clear role for Ca^{2+} in regulating exocytosis in different cell types. Recent evidence from barley aleurone protoplasts suggests the occurrence of both Ca^{2+}-dependent and Ca^{2+}-independent control of exocytosis (Homann and Tester, 1997). In maize coleoptile and root cap cells GTP-binding proteins also appear to play a regulatory role. So far, there are few data on the kinds of proteins involved in the plant cell exocytotic machinery that are regulated by Ca^{2+} and GTP.

Cell volume control

Many plant cells experience changes in their external osmotic environment. The resulting osmotic water fluxes into and out of the cell result in cell volume changes which must be compensated if the cell is to retain turgor (for hyper-osmotic changes) or, in the case of hypo-osmotic changes, prevent cell bursting if turgor pressure exceeds the capacity of the cell wall to resist internal pressure. A plant cell can potentially sense osmotic changes by at least two distinct mechanisms. Changes in mechanical tension exerted on a membrane can potentially be sensed by mechanosensitive receptors. Alternatively the cell may sense changes in the osmotic potential of the cell in response to hydration or dehydration of the cytosol. Evidence for both types of mechanism exists in plant cells. Candidate mechanosensitive receptors include stretch-activated ion channels (SACs) and there is good evidence from patch-clamp studies that these exist on the plasma membrane in plants (Cosgrove and Hedrich, 1991; Ding and Pickard, 1993; Taylor et al., 1996). Changes in vacuolar volume could also potentially be sensed by SACs residing in the tonoplast membrane.

In animal cells there is substantial evidence for the role of SACs in detecting and eliciting responses to cell volume changes, and a regulatory volume decrease in response to hypo-osmotic treatments appears to be mediated by Ca^{2+} to varying degrees in different cell types (McCartey and O'Niel, 1992; Chen et al., 1996). However, the Ca^{2+}-independent signalling of a regulatory cell volume decrease involving protein dephosphorylation had also been observed (Jennings and Schultz, 1991). There is evidence for the involvement of both Ca^{2+} influx and release from intracellular stores in the generation of osmoregulatory Ca^{2+} signals in different animal systems (Hamaza and Okada, 1990; Chen et al., 1996; Clemo and Baumgarten, 1997). Thus in rat pituitary GH3 cells hypo-osmotic treatment produces a relatively long-lasting elevation of Ca^{2+}, accompanied by a regulatory cell volume decrease (Chen et al., 1996). Based on the action of inhibitors of SACs and L-type Ca^{2+} channels a mechanical sensing mechanism has been proposed based on activation of cation-permeable SACs as sensing elements, leading to an elevation of cytosolic Ca^{2+} which also involves the activity of L-type Ca^{2+} channels. Evidence also exists for the involvement of actin in the activation of ion channels involved in cell volume control, based on the action of phalloidin on stretch-activated Cl^- channel activity (Schwiebert et al., 1994) and actin-binding protein (Cantiello et al., 1993). In human epithelial cells, Ca^{2+}-induced Ca^{2+} release is also involved in elevating Ca^{2+} in response to hypo-osmotic treatment (Hamaza and Okada, 1990).

In plant cells evidence for both mechanical and osmosensing mechanisms in cell volume regulation comes from different systems. In the freshwater Charophtye alga *Nitella*, hypo-osmotic treatment induced a transient increase in cytosolic Ca^{2+} which was required for turgor regulation (Tazawa et al., 1994, 1995). Vacuolar perfusion experiments suggested that the trigger for the hypo-osmotic Ca^{2+} increase was cytoplasmic hydration rather than a change in turgor or volume per se. More recently, transient Ca^{2+} elevations have been shown in demembranated cytoplasmic drops from *Nitella* in response to addition of hypotonic solution to the drop in silicone oil (Kikuyama and Tazawa, 1998). Though the mechanism for this osmosensing response is not known, these experiments suggest that generation of the Ca^{2+} signal does not require Ca^{2+} influx and thus is not based on a volume-sensing mechanism in the plasma membrane. Further evidence for the existence of an osmosensing mechanism in plants comes from the work of Lew (1996) with root hairs of *Arabidopsis thaliana*. In these experiments manipulation of cell turgor pressure using a pressure probe electrode was carried out simultaneously with measurements of cell electrical properties. Hypo-osmotic treatment of the cell by changing the external perfusion solution induced a depolarization of approximately +40 mV. In contrast, directly increasing cell turgor by oil injection into the cell does not elicit changes in membrane potential, suggesting that plasma membrane transport is not affected directly by changing cell turgor.

In other plant systems and in yeast there is evidence for the direct involvement of turgor sensing based on mechanical activation of plasma membrane ion channels. In yeast, hypo-osmotic shock induces transient elevations of Ca^{2+} (Batiza et al., 1996). The responses to inhibitors of SACs such as Gd^{3+} and removal of Ca^{2+} from the external solution suggest that the initial rise in Ca^{2+} is generated from intracellular stores but that sustained elevation of Ca^{2+} requires influx through plasma-membrane-located SACs. Zygotes of several types of marine brown algae, such as *Fucus* which grow in the intertidal zone, experience regular and dramatic fluctuations in their external osmotic environment ranging from extreme desiccation to exposure to solution of very low osmotic potential. *Fucus* zygotes are able to regulate their turgor and volume in response to a wide range of osmotic treatments and generate Ca^{2+} transients in response to hypo-osmotic shock (Taylor et al., 1996). The Ca^{2+} transient produced during hypo-osmotic shock is restricted to the rhizoid cell of the two-celled embryo. Moreover, the Ca^{2+} transient initiates at the rhizoid apex and travels as a wave of elevated Ca^{2+} to sub-apical regions, though its progress is restricted to the apical and sub-apical regions of the cell (Fig. 1). Current evidence, based on experiments utilizing the quench of the Ca^{2+} indicator fura-2 by Mn^{2+} as a marker of Ca^{2+} influx, suggests that the Ca^{2+} transient involves Ca^{2+} influx at the point of initiation at the rhizoid apex and release of Ca^{2+} from intracellular stores as it propagates to more sub-apical regions of the cell. The rate of propagation is approximately 5–10 $\mu m\ s^{-1}$, comparable to the propagation velocities of waves of intracellular Ca^{2+} release observed in a variety of animal systems (Jaffe, 1995). No similar physiologically induced Ca^{2+} waves have yet been found in other plant systems, though a wave of intracellular release can be induced in *Papaver* pollen tubes in response to photorelease of caged inositol 1,4,5-trisphosphate [Ins(1,4,5)P_3]. In this case the wave travels towards the growing apex, in the opposite direction to the osmotically induced wave in *Fucus*. While the available evidence suggests a role for SACs in the *Fucus* rhizoid plasma membrane in initiating the Ca^{2+} elevation, the mechanism by which the Ca^{2+} wave is propagated is not clear. This could involve the action of phospholipase C and Ins(1,4,5)P_3 or Ca^{2+}-induced Ca^{2+} release (CICR). In tobacco cells transformed with the Ca^{2+}-dependent photoprotein aequorin, hypo-osmotic shock was accompanied by a transient elevation of cytosolic Ca^{2+}. The magnitude of the Ca^{2+} transient was dependent on the size of the hypo-osmotic shock (Takahashi et al., 1995). The transient showed biphasic behaviour with a slow, small elevation preceding a larger faster transient elevation. Both phases were dependent on the presence of extracellular Ca^{2+} though they had different sensitivity to bafilomycin (a V-type

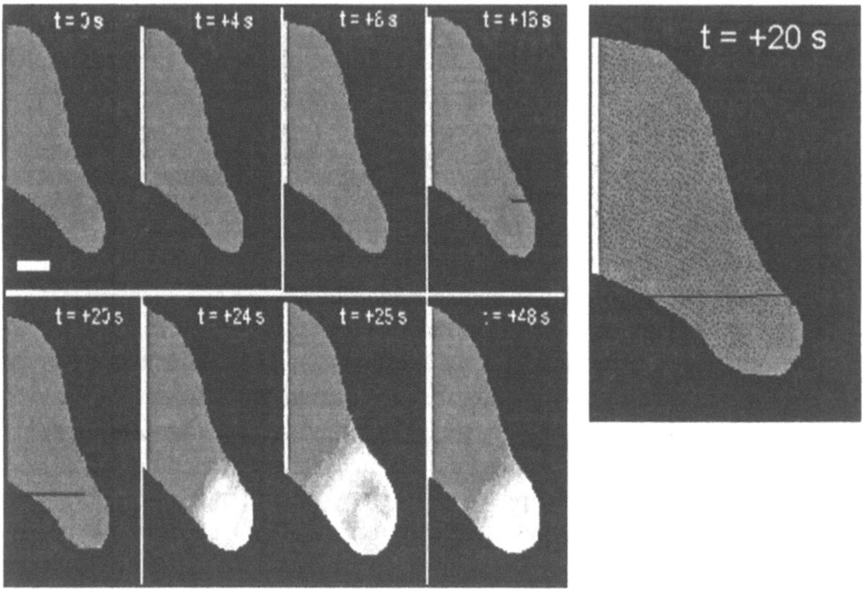

Fig. 1. A Propagation of a Ca^{2+} wave through the apical and sub-apical rhizoid regions of a *Fucus* zygote following hypo-osmotic shock (preincubated for 1 h in sea water + 0.5 M sorbitol and shocked into sea water). Rhizoid cells were microinjected with dextran-conjugated fura-2 and ratio images of cytosolic Ca^{2+} were obtained with a cooled CCD camera as described in Taylor et al. (1996). Transfer to hypo-osmotic solution initiated a wave of cytosolic Ca^{2+} which initiated at the rhizoid apex (❐) and after an initial lag of approximately 20 s spread rapidly from apical to sub-apical regions of the rhizoid. The rapid spread of the calcium signal is shown in the accompanying animation sequence (B)

ATPase inhibitor) and K-252a (a protein phosphorylation inhibitor), suggesting the involvement of different mechanisms for elevating Ca^{2+}.

Transient elevations of Ca^{2+} to hypo-osmotic shock have thus been shown in a higher plant, two unrelated algal species and in yeast. While further studies of other systems are needed, it seems likely that the transient elevation of cytosolic Ca^{2+} may be a general response to hypo-osmotic shock. So far, the spatial characteristics of this signal are only elucidated for the *Fucus* rhizoid, and the mechanisms by which the Ca^{2+} transients are propagated appear to differ between different systems. It is still not clear whether osmosensing involves detection of cell volume changes or a response of intracellular Ca^{2+} to cytoplasmic hydration.

Converging and diverging: downstream responses of osmoregulation and polarity

Solute fluxes and membrane cycling

Early downstream responses to osmotic stimuli include the regulation of ion channels which facilitate ion and water fluxes across the plasma membrane, regulation of membrane turnover, and, further downstream, regulation of the expression of genes involved in osmosensing and regulation. Responses observable during or soon after the osmoregulatory Ca^{2+} signal in *Fucus* include solute loss, indicating the activation of ion efflux channels (Taylor et al., 1996). During normal growth the membrane potential of the *Fucus* embryo is dominated by a high conductance to K^{+} (Taylor and Brownlee, 1993) Hypo-osmotic treatment causes a transient depolarization (Fig. 3) which is consistent with opening of cation-selective channels as described previously (Taylor et al., 1996) or the activation of channels allowing Cl^{-+} efflux. The time course of the transient depolarization closely mirrors that of the transient elevation of cytosolic Ca^{2+} (Fig. 2). Since the plasma membrane can be considered to have a very limited capacity for expansion, increased membrane surface area must be accommodated by incorporation of new membrane. Freeze fracture electron microscopy studies of *Pelvetia* zygotes have shown increased exocytotic vesicle incorporation into the plasma membrane following hypo-osmotic shock (Gilkey and Staehelin, 1989). It was proposed that these vesicles can increase the number of Cl^{-} efflux channels in the plasma membrane, allowing increased solute loss. Further evidence for increased plasma membrane turnover in the *Fucus* rhizoid apex in response to osmotic shock comes from experiments with the fluorescent membrane probe FM1–43. This probe incorporates into and fluorescently labels the plasma membrane (Smith and Betz, 1996). *Fucus* zygotes incubated in the presence of FM1–43 showed an immediate increase in fluorescence associated with the plasma membrane followed by a slow internalization of the dye at the rhizoid apex (Fig. 3). Hypo-osmotic treatment (from sea water into 50% sea water) resulted, after an initial lag of several seconds, in a dramatic increase in dye internalization at the rhizoid apex, a response which we interpret as increased membrane retrieval. The time course for increased membrane internalization shows that this occurred predominantly after the Ca^{2+} transient and membrane depolarization and corresponded with the regulatory cell volume decrease (Fig. 2). It can thus be concluded that increased exocytosis associated with cell swelling is followed by increased membrane retrieval as cell volume returns to normal.

In barley aleurone protoplasts, application of small positive hydrostatic pressure via a patch pipette in the whole-cell recording configuration caused an increase in cell surface area, monitored as an increase in whole-cell capacitance (Zorec and Tester, 1993). The increase in cell surface area was far greater than that induced by elevated Ca^{2+} in the same cells by intracellular perfu-

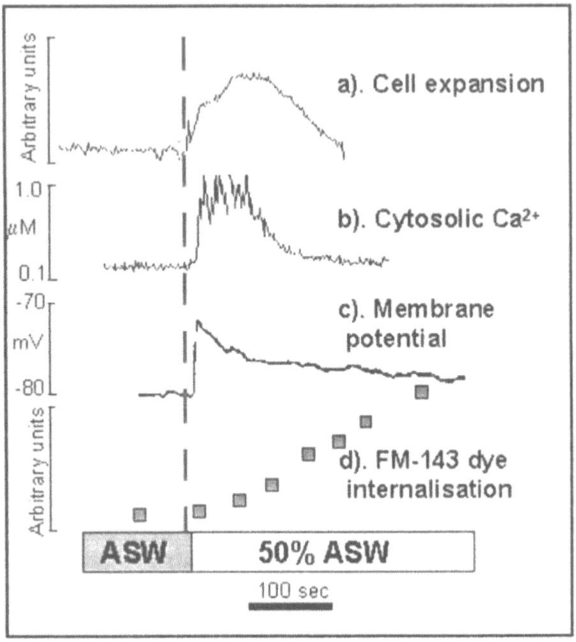

Fig. 2. Temporal correlation of cell expansion, cytosolic Ca^{2+} transient, membrane potential depolarization and internalization of the membrane dye FM1-43 in the *Fucus* rhizoid apex following a hypo-osmotic shock. Hypo-osmotic shock induced a reversible increase in cell volume lasting approximately 200 s. Cytosolic Ca^{2+} increased and membrane potential depolarized rapidly and transiently following an initial lag of several seconds. Internalization of the membrane dye FM1-43 (see Fig. 3) occurred more gradually and continued during the regulatory cell volume decrease, recovery of cytosolic Ca^{2+} levels and repolarization of the membrane potential. Cell expansion was monitored photometrically as described in Taylor et al. (1996). Cytosolic Ca^{2+} was monitored ratiometrically with the following microinjection of fura-2 dextran (Taylor et al., 1996). Membrane potential was recorded with an intracellular microelectrode (Taylor and Brownlee, 1993)

sion of Ca^{2+} buffer via the patch pipette. The pressure-induced cell surface increase was rapidly reversible on release of pressure, indicating that membrane retrieval can occur at a rapid rate.

MAP kinases, osmosensing and polarity

Osmotic sensing in yeast involves activation of a mitogen-activated protein (MAP) kinase pathway. In yeast the HOG signalling pathway, involving *HOG1* and *PBS2* genes (Schuller et al., 1994), has been shown to link responses to osmolarity and cell cycle regulation via osmotic-stress-induced activation of a MAP kinase homologue by wis1 kinase, promoting the onset of mitosis (Shiozaki and Russell, 1995). Wis1 (a MAP kinase kinase) has also been shown to control the synthesis of osmo-inducible proteins such as glycerol-3-phos-

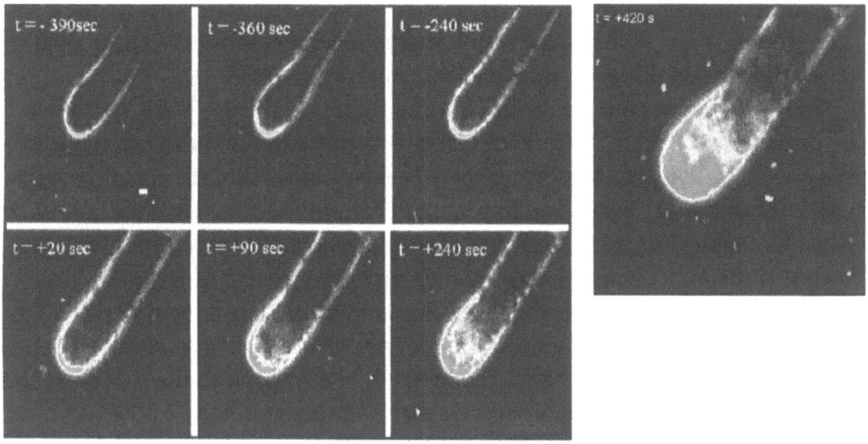

Fig. 3. Confocal images of the pattern of internalization of the fluorescent membrane dye FM1-43 in a growing rhizoid cell of *Fucus serratus*. Embryos were incubated in 1 μM FM1-43 for the duration of the experiment. Hypo-osmotic treatment (from sea water to 50% sea water) was given at *t*=0. Little or no dye internalization was observed for at least 390 s prior to the hypo-osmotic shock. However, dye internalization was apparent within 20 s of the shock and continued for the duration of the recording (240 s)

phate dehydrogenase. Evidence is also accumulating for a role for MAP kinase signalling in higher plants, particularly in response to mechanical, osmotic and wounding stimuli (Bogre et al., 1996; Machida et al., 1997). A MAP kinase homologue from *Pisum sativum* (*PSMAPK*) can rescue HOG1 mutants under conditions of osmotic stress (Popping et al., 1996). A MAP kinase has also been implicated in cold and drought signalling in *Alfalfa* (Jonak et al., 1996).

In yeast osmotic sensing pathways and polarity are closely linked and involve shared signalling components. In budding yeast increased external osmolarity involves growth arrest and loss of actin filaments from the bud (Brewster and Gustin, 1994). Recovery involves re-establishment of actin at the former bud site and resumption of growth. Mutants lacking the MAP kinase encoded by *HOG1* are unable to re-establish actin filaments at the original growth site during recovery. Actin filaments and the growth site re-localized randomly in the mutants, suggesting that the HOG1 pathway has a direct role in determining the positioning of the polar axis. Activation of *SLT2* (*MPK1*) MAP kinase, which is involved in responses to hypo-osmotic shock, has also been shown during polarized cell growth in *Saccharomyses cereviseae* (Zarzov et al., 1996).

In higher plants we know less about the pathways leading to altered gene expression in response to polarizing stimuli. However, during root development, particular epidermal cells to begin polarize and adopt a hair phenotype in response to the position and presence of the plant growth regulator ethylene (Dolan and Roberts, 1995; Tanimoto et al., 1995). Interestingly the ethylene signal transduction pathway is a two-component pathway (Chang and Meyerowitz, 1995; Fluhr, 1998) which shares several common features with the yeast HOG1 MAP kinase cascade (Posas and Saito, 1998).

In higher plant embryos an increasing number of genes are now known to be directly involved in patterning (Laux and Jurgens, 1997) though so far only the *gnom/EMB30* gene has been shown to have a direct role in apical-basal polarity (Mayer et al., 1993). The product of this gene shows sequence homology with the yeast proteins GEA2P and GEA1p which are involved in intracellular vesicle transport, raising the possibility that *GNOM* may function in targeted exocytosis which has been shown in *Fucus* to be required for polar axis fixation (Shaw and Quatrano, 1996). Different patterns of Ca^{2+} signal can influence gene expression in animal cells (Dolmetsch et al., 1997). While it is clear that interactions occur between osmotic Ca^{2+} signals and gene expression, it is much less clear whether the expression of genes involved in polarization is affected by localized Ca^{2+} signals associated with polarization. In *Fucus* zygotes, while there is good evidence that Ca^{2+} signalling is essential for polarization and orientation responses (Speksnijder et al., 1989; Robinson, 1996; Love et al., 1997), there is currently little direct evidence that the apical Ca^{2+} gradient is directly required for determination of cell fate, though it seems reasonable to assume that targeted exocytosis, which is required for polar axis fixation and rhizoid cell fate determination, is controlled by the apical Ca^{2+} gradient. While buffering cytosolic Ca^{2+} leads to prevention of rhizoid cell formation, it is still not clear whether this reflects inhibition of rhizoid cell growth or rhizoid cell determination. The restriction of the polarization-associated Ca^{2+} signal to the rhizoid apex means that it is spatially separated from the nucleus and is unlikely to affect gene expression directly. However, it is not yet clear whether Ca^{2+} transients and waves in response to osmotic treatments impinge on the nucleus.

Concluding remarks

The above examples indicate that a variety of different signal-response coupling pathways share common components at the level of both mobile messengers such as Ca^{2+} and phosphorylation cascades such as MAP kinase cascades. The answer to the question of how specificity of signalling can be achieved under these circumstances lies in more detailed knowledge of the interactions between different components and the spatial and temporal patterns of their activation.

References

Batiza, A.F., Schultz, T., Masson, P. (1996) Yeast respond to hypotonic shock with a calcium pulse. *J. Biol. Chem.* **271**:23357–23362

Benkert, R., Obermeyer, G., Bentrup, F.-W. (1997) The turgor pressure of growing lily pollen tubes. *Protoplasma* **198**:1–8

Berger, F. Brownlee, C. (1993) Ratio confocal imaging of free cytoplasmic calcium gradients in polarizing and polarized *Fucus* zygotes. *Zygote* **1**:9–15

Berger, F., Taylor, A.R., Brownlee, C. (1994) Cell fate determination by the cell wall in early *Fucus* development. *Science* 263:1421–1423

Bogre, L., Ligterink, W., Heberlebors, E., Hirt, H. (1996) Mechanosensors in plants. *Nature* **383**:489–490

Bouget, F.-Y., Berger, F., Brownlee, C. (1998) Position-dependent control of cell fate in the *Fucus* embryo: role of intercellular communication. *Development* **125**:1999–2008

Brawley, S.H., Robinson, K.R. (1985) Cytochalasin treatment disrupts the endogenous currents associated with cell polarization in fucoid zygotes. *J. Cell Biol.* **100**:1173–1184

Brewster, J.L., Gustin, M.C. (1994) Positioning of cell growth and division after osmotic stress requires a MAP kinase pathway. *Yeast* **10**:425–439

Brownlee, C., Pulsford, A.L. (1988) Visualization of the cytoplasmic Ca^{2+} gradient in *Fucus serratus* rhizoids: correlation with cell ultrastructure and polarity. *J. Cell Sci.* **91**:249–256

Brownlee, C., Wood, J.W. (1986) a gradient of free cytosolic calcium in rhizoid cells of *Fucus serratus*. *Nature* **320**:624–626

Cantiello, H.F., Prat, A.G., Bonventre, J.V., Cunningham, C.C., Hartwig, J.H., Ausiello, D.A. (1993) Actin-binding protein contributes to cell volume regulatory ion channel activation in melanoma cells. *J. Biol. Chem.* **268**:4596–4599

Carroll, A.D., Moyen, C., van Kesteren, P., Tooke, F., Battey, N.H., Brownlee, C. (1998) Ca^{2+}, annexins and GTP modulate exocytosis from maize root cap protoplasts. *Plant Cell* **10**:1267–1276

Chang, C., Meyerowitz, E. (1995) The ethylene hormone response in *Arabidopsis*: a eukaryote two-component signalling system. *Proc. Natl. Acad. Sci. USA* **92**:4129–4133

Chen, Y., Simasko, S.M., Niggel, J., Sigurdson, W.J., Sachs, F. (1996) Ca^{2+} uptake in GH3 cells during hypotonic swelling: the sensory role of stretch-activated ion channels. *Am. J. Physiol.* **270**:C1790–C1798

Clemo, H.F., Baumgarten, C.M. (1997) Swelling-activated Gd^{3+}-sensitive cation current and cell volume regulation in rabbit ventricular myocytes. *J. Gen. Physiol.* **110**:297–312

Cosgrove, D. (1993) Wall extensibility: its nature measurement and relationship to plant cell growth. *New Phytol.* **124**:1–23

Cosgrove, D.J., Heidrich, R. (1991) Stretch-activated Cl^-, K^+ and Ca^{2+} channels coexisting in plasma membrane of guard cells of *Vicia faba* L. *Planta* **186**:143–153

Ding, P.J., Pickard, B. (1993) Mechanosensory Ca^{2+}-selective cation channels in epidermal cells. *Plant J.* **3**:83–110

Dolan, L., Roberts, K. (1995) The development of cell pattern in the root epidermis. *Philos. Trans. R. Soc.* **350**:95–99

Dolmetsch, R.E., Lewis, R.S., Goodnow, C.C., Healy, J.I. (1997) Differential activation of transcription factors induced by Ca^{2+} response amplitude and duration. *Nature* **386**:855–858

Felle, H.H., Hepler, P.K. (1997) The cytosolic Ca^{2+} concentration gradient of *Sinapis alba* root airs as revealed by Ca-selective microelectrode tests and fura-dextran ratio imaging. *Plant Physiol.* **114**:39–45

Fluhr, R. (1998) Ethylene perception: from two-component signal transducers to gene induction. *Trends Plant. Sci.* **3**:141–146

Franklin-Tong, V.E., Hackett, G., Hepler, P.K. (1997) Ratio-imaging of Ca^{2+} in the self-incompatibility response of *Papaver rhoeas. Plant J.* **12**:1375–1386

Geitman, A., Li, Y.Q., Cresti, M. (1996) The role of the cytoskeleton and dictyosome activity in the pulsatory growth of *Nicotiana tabacum* and *Petunia hybrida* pollen tubes. *Bot. Acta* **109**:102–109

Gilkey, J.C., Staehelin, L.A. (1989) A new organelle related to osmoregulation in ultrarapidly frozen *Pelvetia* embryos. *Planta* **178**:425–435

Hamaza, A., Okada, Y. (1990) Involvement of Ca^{2+}-induced Ca^{2+} release in the volume regulation of human epithelial cells exposed to a hypotonic medium. *Biochem. Biophys. Res. Commun.* **167**:287–293

Harold, F.M., Harold, R.L., Money, N.P. (1995) What forces drive cell wall expansion? *Can. J. Bot.* **73**:379–383

Hermann, A., Felle, H. H. (1995) Tip growth in root hair cells of *Sinapis alba* L: significance of internal and external Ca^{2+} and H^+. *New Phytol.* **129**:523–533

Hohl, M., Schopfer, P. (1992) Growth at reduced turgor: irreversible and reversible cell wall extension of maize coleoptiles and its implications for the theory of cell growth. *Planta* **187**:209–217

Holdaway-Clarke, T.l., Feijo, J.A., Hackett, G.R., Kunkel, J.G., Hepler, P.K. (1997) Pollen tube growth and the intracellular cytosolic calcium gradient oscillate in phase while extracellular calcium influx is delayed. *Plant Cell* **9**:1999–2010

Homann, U., Tester, M. (1997) Ca^{2+}-independent and Ca^{2+}-GTP binding protein-controlled exocytosis in a plant cell. *Proc. Natl. Acad. Sci. USA* **94**:6565–6570

Jaffe, L.F. (1995) Calcium waves and development. *CIBA Found. Symp.* **188**:4–12

Jennings, M.L., Schultz, R.K. (1991) Okadaic acid inhibition of KCl cotransport: evidence that protein dephosphorylation is necessary for activation of transport by either cell swelling or *N*-ethylmaleimide. *J. Gen. Physiol.* **97**:799–818

Jonak, C., Kiegerl, S., Ligterink, W., Barker, P.J., Huskinsson, N.S., Hirt, H. (1996) Stress signalling in plant: a mitogen-activated protein kinase pathway is activated by cold and drought. *Proc. Natl. Acad. Sci. USA* **93**:11274–11279

Kikuyama, M., Tazawa, T. (1998) Involvement of water channels in the hydration-induced calcium release in *Nitella flexilis*. *Plant Cell Physiol.* **39**:830–835

Kropf, D.L. (1997) Induction of polarity in fucoid zygotes. *Plant Cell* **9**:1011–1020

Kropf, D.L., Kloareg, B., Quatrano, R.S. (1988) Cell wall is required for fixation of the embryonic axis in *Fucus* zygotes. *Science* **239**:187–190

Kutschera, U. (1996) Cessation of cell elongation in rye is accompanied by a loss of cell wall plasticity. *J. Exp. Bot.* **47**:1387–1349

Kutschera, U., Kohler, K. (1994) Cell elongation, turgor and osmotic pressure in developing sunflower hypocotyls. *J. Exp. Bot.* **45**:591–595

Laux, T., Jurgens, G. (1997) Embryogenesis: a new start in life. *Plant Cell* **9**:989–1000

Lew, R. (1996) Pressure regulation of the electrical properties of growing *Arabidospis thaliana* L. root hairs. *Plant Physiol.* **112**:1089–1100

Love, J., Brownlee, C., Trewavas, A.J. (1997) Ca^{2+} and calmodulin dynamics during photopolarization in *Fucus serratus* zygotes. *Plant Physiol.* **115**:249–261

Machida, Y., Nishihama, R., Kitakura, S. (1997) Progress in studies of plant homologues of mitogen-activated protein (MAP) kinases and potential upstream components in kinase cascades. *Crit. Rev. Plant Sci.* **16**:481–496

Malho, R., Trewavas, A.J. (1996) Localized apical increases of cytosolic free calcium control pollen tube orientation. *Plant Cell* **8**:1935–1949

Mayer, U., Büttner, G., Jürgens, G. (1993) Apical-basal pattern formation in the *Arabidopsis* embryo: studies on the role of the gnom gene. *Development* **117**:149–162

McAinsh, M.R., Hetherington, A.M. (1998) Encoding specificity in Ca^{2+} signalling systems. *Trends Plant Sci.* **3**:32–36

McAinsh, M.R., Brownlee, C., Hetherington, A.M. (1997) Calcium ions as second messengers in guard cell signal transduction. *Physiol Plant* **100**:16–29

McCabe, P.F., Valentine, T.A., Forsberg, L.S., Pennell, R.I. (1997) Soluble signals from cells identified at the cell wall establish a developmental pathway in carrot. *Plant Cell* **9**:2225–2241

McCarty, N.A., O'Neil, R.G. (1992) Calcium signalling in cell volume regulation. *Physiol Rev* **72**:1037–1061

Messerli, M., Robinson, K.R. (1997) Tip localized Ca^{2+} pulses are coincident with peak pulsatile growth rates in pollen tubes of *Lilium longiflorum*. *J. Cell Sci.* **110**:1269–1278
Miller, D., Callaham, D.A., Gross, D.J., Hepler, P.K. (1992) Free Ca^{2+} gradient in growing pollen tubes of *Lilium*. *J. Cell Sci.* **101**:7–12
Money, N.P., Harold, F.M. (1992) Extension growth of the water mold *Achlya*: interplay of turgor and wall strength. *Proc. Natl. Acad. Sci. USA* **89**:4245–4249
Money, N.P., Harold, F.M. (1993) Two water molds can grow without measurable turgor pressure. *Planta* **190**:426–430
Pierson, E.S., Miller, D.D., Callaham, D.A., Shipley, A.M., Rivers, B.A., Cresti, M., Hepler, P.K. (1994) Pollen tube growth is coupled to the extracellular calcium ion flux and the intracellular calcium gradient: effect of BAPTA-type buffers and hypertonic media. *Plant Cell* **6**:1815–1828
Popping, B., Gibbons, T., Watson, M.D. (1996) The *Pisum sativum* MAP kinase homologue (PSMAPK) rescues the *Saccharomyces cerevisiae* HOG1 deletion mutant under conditions of high osmotic stress. *Plant Mol. Biol.* **31**:355–363
Posas, F., Saito, H. (1998) Activation of the yeast SSK2 MAP kinase kinase by the SSK1 two component response regulator. *EMBO J.* **17**:1385–1394
Pritchard, J. (1994) The control of cell expansion in roots. *New Phytol.* **127**:3–26
Rathore, K.S., Cork, R.J., Robinson, K.R. (1991) A cytoplasmic gradient of Ca^{2+} is correlated with the growth of lily pollen tubes. *Dev. Biol.* **148**:612–619
Robinson, K.R. (1996) Calcium and the photopolarization of *Pelvetia* zygotes. *Planta* **198**:378–384
Schuller, C., Brewster, J.L., Alexander, M.R., Gustin, M.C., Ruis, H. (1994) The HOG pathway controls osmotic regulation of transcription via the stress response element (STRE) of the *Saccharomyces cerevisiae* CIT1 gene. *EMBO J.* **13**:4382–4389
Schwiebert, E.M., Mills, J.W., Stanton, B.A. (1994) Actin-based cytoskeleton regulates a chloride channel and cell volume in a renal cortical collecting duct cell line. *J. Biol. Chem.* **269**:7081–7089
Shaw, S.L., Quatrano, R.S. (1996) The role of targeted secretion in the establishment of cell polarity and the orientation of the division plane in *Fucus* zygotes. *Development* **122**:2623–2630
Shiozaki, K., Russell, P. (1995) Cell cycle control linked to extracellular environment by MAP kinase pathway in fission yeast. *Nature* **378**:739–743
Smith, C.B., Betz, W.J. (1996) Simultaneous independent measurement of endocytosis and exocytosis. *Nature* **380**:531–534
Speksnijder, J.A., Miller, A.L., Weisenseel, M.H., Chen, T.-H., Jaffe, L.F. (1989) Calcium buffer injections block fucoid egg development by facilitating calcium diffusion. *Proc. Natl. Acad. Sci. USA* **86**:6607–6611
Steer, M., Steer, J.M. (1989) Pollen tube growth. *New Phytol.* **111**:323–358

Takahashi, K., Isobe, M., Kniught, M.R., Trewavas, A.J., Muto, S. (1995) Hypo-osmotic shock induces increases in cytosolic Ca^{2+} in tobacco suspension culture cells. *Plant Physiol.* **113**:587–594
Tanimoto, M., Roberts, K., Dolan, L. (1995) Ethylene is a positive regulator of root hair development in *Arabidopsis thaliana. Plant J.* **8**:943–948
Taylor, A.R., Brownlee, C. (1993) Calcium and potassium currents in the *Fucus* egg. *Planta* **189**:109–119
Taylor, A.R., Manison, N.F.H., Fernandez, C., Wood, J.W., Brownlee, C. (1996) Spatial organization of calcium signalling involved in cell volume control in the *Fucus* rhizoid. *Plant Cell* **8**:2015–2031
Tazawa, M., Yoko-o, T., Mimura, T., Kikuyama, M. (1994) Intracellular mobilization of Ca^{2+} and inhibition of cytoplasmic streaming induced by transcellular osmosis in internodal cells of *Nitella flexilis. Plant Cell Physiol.* **35**:63–72
Tazawa, M., Shimada, K., Kikuyama, M. (1995) Cytoplasmic hydration triggers a transient increase in cytoplasmic Ca^{2+} concentration in *Nitella flexilis. Plant Cell Physiol.* **36**:335–340
Thiel, G., Rupnik, M., Zorec R. (1994) Raising the cytosolic Ca^{2+} concentration increases the membrane capacitance of maize coleoptile protoplasts: evidence for Ca^{2+}-stimulated exocytosis. *Planta* **195**:305–308
Tomos, A.D., Pritchard, J. (1994) Biophysical and biochemical control of cell expansion in roots and leaves. *J. Exp. Bot.* **45**:1721–1731
Van den Berg, C., Willemsen, V., Hage, W., Weisbeek, P., Scheres, B. (1995) Cell fate in the *Arabidopsis* root meristem determined by directional signalling. *Nature* **378**:62–65
Wymer, C.L., Bibikova, T.N., Gilroy, S. (1997) Cytoplasmic free calcium distributions during the development of root hairs of *Arabidopsis thaliana. Plant J.* **12**:427–439
Zarzov, P., Mazzoni, C., Mann, C. (1996) The SLT2(MPK1) MAP kinase is activated during periods of polarized cell growth in yeast. *EMBO J.* **15**:83–91
Zhu, G.L., Boyer, J.S. (1992) Enlargement in *Chara* studied with a turgor clamp: growth rate is not determined by turgor. *Plant Physiol.* **100**:2071–2080
Zorec, R., Tester, M. (1992) Cytoplasmic Ca^{2+} stimulates exocytosis in a plant secretory cell. *Biophys. J.* **63**:864–867
Zorec, R., Tester, M. (1993) Rapid pressure driven exocytosis-endocytosis cycle in a single plant cell. *FEBS Lett.* **333**:283–286

Do single mitochondria contain zones with different membrane potential?

J. Bereiter-Hahn · M. Vöth

Cinematic Cell Research Group, Biocentre, Johann Wolfgang Goethe University Frankfurt/M., Germany

Received: 23 July 1998 / Accepted: 2 November 1998 / Published: 11 November 1998

Abstract. Observations of Lan Bo Chen's group using a mitochondria-selective fluorochrome 5,5',6,6'- tetrachloro- 1,1',3,3'- tetraethylbenzimidazolocarbocyanine iodide (JC-1) indicate that mitochondria *in situ* may have zones of different electrochemical potential along their length. This was indicated by the formation of J-aggregates of this dye at distinct sites along a single mitochondrion. Also, intensity variations along single mitochondria were found with diamino-styryl-pyridinium methiodide (DASPMI), another fluorochrome that selectively stains mitochondria depending on their electrochemical potential. DASPMI exchanges easily with the cytoplasm and changes its quantum yield when bound to mitochondrial membranes. Therefore, fluorescence intensity is primarily controlled by the membrane environment rather than by mass accumulation. Two possible explanations of intramitochondrial fluorescence intensity variations have to be discussed: variations in the amount of mitochondrial inner membrane per unit of projection area (or voxel), and differences in the electrochemical gradient. This problem has been approached by comparing fluoro-micrographs of mitochondria in endothelial cells stained with either JC-1 or DASPMI with electron micrographs of the same mitochondria after fixation with glutardialdehyde and osmium tetroxide and ultrathin sectioning. JC-1 red fluorescence (revealing J-aggregate formation) as well as high-intensity staining with DASPMI correlate roughly with the local thickness of mitochondria; no differences in the crista organization are revealed for those areas or mitochondria exhibiting red JC-1 fluorescence and those with green fluorescence. The distance between red fluorescing areas in a single mitochondrion seem to be caused by competition for dye molecules placed in between centres of JC-1 aggregation. Isolated mitochondria are of uniform small size and spherical shape; therefore, no differ-

Correspondence to: J. Bereiter-Hahn (e-mail: bereiter-hahn@zoology.uni-frankfurt.de, Fax: +49-69-79829607)

ences in shape interfere with JC-1 staining. Thus JC-1 may be an appropriate indicator of membrane potential in isolated mitochondria. In living cells mitochondria often are large and elongated, and thus the situation is not straightforward to interpret. However, evidence is provided that there are submitochondrial zones, which differ in membrane potential from one adjacent area to another, because DASPMI staining of intramitochondrial zones reveals differences in fluorescence intensity and preferred photodamage of these areas. In some cases separation of the zones of higher membrane potential by cristae traversing the whole diameter of a mitochondrion has been observed. Local photobleaching of stained mitochondria results in a loss of fluorescence along the total length of a mitochondrion. However, this type of bleaching develops over tens of seconds, not in the very short time range (e.g. ms) expected from the discharge of all the membranes if they were electrically coupled.

Key words. Cell cultures · DASPMI · Fluorescence · JC-1 · Membrane potential Mitochondria

Introduction

Fluorescent probes have been applied widely as optical indicators of the membrane potential in several types of cell, isolated organelles and lipid vesicles (e.g. Bereiter-Hahn, 1990; Chen, 1988; Petit, 1992; Walsh Kinnally *et al.*, 1978). The technique relies on membrane-potential-dependent partitioning of charged lipophilic dye molecules across the membrane. Changes in membrane potential result in changes in the intensity of dye fluorescence which have been termed redistribution signals by Cohen and Salzberg (1978). In studies of isolated membranes and organelles, a large range of probes has been used for a long time (Walsh Kinnally *et al.*, 1978). However, the evaluation of mitochondrial membrane potential in living cells requires the selective staining of these organelles by the fluorochrome, and interference with the potential across the plasma membrane may influence these measurements. Selective fluorochroming of mitochondria was first reported for rhodamine applied to plant cells (Johannes, 1941), and it was re-detected and attributed to a specific rhodamine isoform, rhodamine 123 (Rh123), by Lan Bo Chen's group in 1978 (Johnson *et al.*, 1980). This fluorochrome is widely used now to visualize mitochondria and evaluate their membrane potential. In 1976, another fluorochrome, dimethyl-aminoystyryl-propidium-methiodide (DASPMI), was described (Bereiter-Hahn, 1976) which selectively accumulates in the mitochondria of a variety of cells (Eng *et al.*, 1989; Fuchs *et al.*, 1987; Irion *et al.*, 1993; Leip *et al.*, 1994; McCormick, 1990).

In parallel, fluorochromes have been developed to allow detection of respiratory activity by oxidation-induced fluorescence. Among those, dihydrorhodamine (Sobreira *et al.*, 1996) and Mitotracker (Molecular Probes,

Eugene, Ore., USA) have become the most important. The achievements in the field of mitochondria fluorochromation are summarized by Chen (1989) and Bereiter-Hahn (1990).

One of the most interesting dyes in the series of mitochondria-selective fluorochromes is 5,5',6,6'-tetrachloro-1,1',3,3'-tetraethylbenzimidazolocarbocyanine iodide (JC-1) (Reers *et al.*, 1991; Smiley *et al.*, 1991). JC-1 appears in two forms: the J-aggregates, which emit red fluorescence, and the monomolecular form, which emits green fluorescence. The green fluorescence selectively marks mitochondria, and the red fluorescence intensity is linearly related to the membrane potential of isolated mitochondria. In some very long mitochondria a mosaic staining pattern, red fluorescing zones separated by green fluorescing spaces, has been detected (Smiley *et al.*, 1991). This staining pattern seems to indicate membrane potential variations along a single mitochondrion.

Two earlier findings are not consistent with this interpretation: (1) the high mobility of charges along membranes (Hackenbrock, 1981) should prevent any potential differences along single mitochondria, and (2) the membrane potential breaks down along the whole length of a mitochondrion when it is permeabilized locally by strong irradiation (Amchenkova *et al.*, 1988). Both these observations assume a low internal resistance along a mitochondrion. Whether this low resistance instantaneously dissipates local potential differences depends on the equilibrium between local proton transport and proton flux along the length of a mitochondrion. Little is known about these parameters at the level of a single mitochondrion.

The energy status of mitochondria is also associated with the prevalence of certain fine structural details.

The morphology of mitochondria *in situ* has been studied under carefully controlled metabolic conditions in several cells, including Ehrlich ascites tumor cells by Hackenbrock *et al.* (1971) and by Koukl *et al.* (1977), bovine adrenal cells (Allmann *et al.*, 1970 a, b), and cardiomyocytes (Harris *et al.*, 1965, 1968, 1969; Koke *et al.*, 1976), and more recently in amphibian endothelial cells (Bereiter-Hahn and Vöth, 1983). Most of our knowledge, however, has been gained by investigating isolated mitochondria and following their morphological reactions by such methods as light scattering and electron microscopy (for review see Munn, 1974).

The correlation of metabolic state with morphological changes has long been a matter of discussion (Williams *et al.*, 1970). Since the early studies of Hackenbrock (Hackenbrock, 1968; Hackenbrock *et al.*, 1971) and of Green and his group (Allmann *et al.*, 1970a, b; Harris *et al.*, 1965, 1969; Penniston *et al.*, 1968), it has been agreed that the condensed conformation represents the low-energy state and the orthodox conformation the high-energy state. „Low-energy state“ corresponds to state 3 respiration (according to Chance and Williams, 1956) and to mitochondria with inhibited electron transport. „High-energy state“ corresponds to state 4 respiration.

Further confirmation of this interpretation came from ADP and ATP injection studies. Single endothelial cells and single amoebae have been injected with ADP or ATP and then the changes of mitochondrial fine structure revealed using electron microscopy (Bereiter-Hahn and Vöth, 1983, 1994; Ord and Smith, 1982). In both these species, mitochondria with the „classic" crista pattern and matrix of low electron density (orthodox conformation) were found after injection of ATP. An increase of intracellular ADP caused the mitochondria to „contract", the matrix to become electron dense, and the cristae to turn to a vesicular form through swelling of the intracristal compartment („condensed conformation"). Mitochondria in untreated cells achieved an inner compartment of intermediate structure. Although the membrane potential of these mitochondria has not been measured, a high electrochemical potential can be assumed for the ATP-treated organelles and a lower electrochemical potential can be attributed to the mitochondria exposed to ADP.

As a consequence, if pronounced variations in the electrochemical potential can occur along mitochondria these should be accompanied by local differences in the appearance of the inner mitochondrial compartment, i.e. differences in the degree of condensation. This principle was used to investigate whether JC-1 red fluorescence can be correlated to a specific state of mitochondrial inner compartment condensation. In addition, the intensity distribution of another fluorochrome, DASPMI, which also monitors membrane potential, has been related to mitochondrial fine structure.

Materials and methods

The experiments were performed using a line of endothelial cells (XTH-2) derived from *Xenopus laevis* tadpole hearts (Schlage *et al.*, 1981). These cells can be kept at room temperature (approx. 24°C) for the experiments, they are very flat and thus allow the observation of single mitochondria by phase-contrast and fluorescence microscopy. For the experiments the cells were seeded on cover glasses (24 mm diameter) and used at preconfluent density.

DASPMI staining

DASPMI (Mol. Wt. 366) was added to the cultures from an aqueous stock solution (5×10^{-4} mol/l) to the cultures at a final concentration of 5×10^{-6} mol/l. Observation of the living cultures started after 60 min of incubation, and rinsing the cultures with DASPMI-free culture medium was done using the filter set for fluorescein-isothiocyanate (FITC) on a Leica confocal microscope TCS based on the inverted microscope Leica DMB equipped with a 100× objective lens. Pinholes were opened to maximal size to allow imaging of the whole thickness of the thin peripheral cytoplasm within a single scan (to reduce bleaching effects of the very sensitive living cells).

JC-1 staining

JC-1 was a generous gift from Professor Lan Bo Chen (Boston). It was dissolved in dimethylsulphoxide (DMSO) to make a stock solution (1 mg/ml). For preparation of the experimental solution this stock solution was diluted 1:100 with the saline of amphibian culture medium plus 10% fetal calf serum (pH 7.6) and stirred at 37°C for 20 min. After filtration of this solution through a Millipore filter (0.2 μm pore size), 132 μl was added to 2 ml of culture medium. Cells were incubated for 10 min at room temperature in this final dye solution and then observed using the fluorescence filter set for FITC on a Zeiss Axiovert inverted microscope, objective lens 100×.

Fixation for electron microscopy

After observation of the fluorescence the cultures were immediately fixed in 3% glutaraldehyde (in 0.05 mol/l phosphate buffer, pH 7.0) for 1 h, then postfixed in 1% osmium tetroxide (in water) for 1 h, dehydrated in acetone and embedded in epon 812.

Results

DASPMI – fluorescence of single mitochondria

DASPMI is a weak cationic dye which emits constant fluorescence in water at a pH exceeding 4.0. On changing the environment from the aqueous phase to a lipid phase, the quantum yield increases up to 12-fold. Uptake of this dye into mitochondria can be recorded immediately because of an immediate increase in fluorescence intensity. No comparable change in fluorescence occurs in suspensions of mitochondria in the presence of rhodamine 6GO. We believe that the reason for this different behaviour is where the dye molecules accumulate in a membrane-potential-dependent manner. DASPMI is thought to remain in the inner membrane, a hydrophobic environment, while rhodamines accumulate in the matrix space, a hydrophilic environment. Therefore, the latter are not subjected to changes in their immediate vicinity causing alterations in quantum yield while moving from the cytoplasm into mitochondria. The dyes accumulate according to a Nernst distribution and may be retained for a long time inside the mitochondria (Van Itallie *et al.*, 1990). DASPMI rapidly exchanges between the mitochondrial environment and the solvent in the test tube, or the cytoplasm in living cells. This interpretation is supported by the fact that DASPMI has a high affinity for hydrophobic environments (at 10 μmol/l the partition coefficient between water and chloroform is 1:62). The net positive charge of the dye is one factor responsible for its preferred targeting to mitochondrial membranes.

The relationship between energization of mitochondria and DASPMI fluorescence has been determined in mitochondrial suspensions (Bereiter-Hahn, 1976), in living cells (Bereiter-Hahn *et al.*, 1983) and in the beating heart (Fuchs *et al.*, 1987; Thürich *et al.*, 1998). A direct relationship with the mitochondrial membrane potential according to the Nernst equation exists. This has been shown by varying the K^+ distribution potential in mitochondrial suspensions treated with nigericin.

For large numbers of mitochondria, their fluorescence corresponds to their membrane potential if related to the total mass of mitochondrial protein. But, what about the fluorescence intensity of single mitochondria and the fluorescence distribution along a single mitochondrion? In living cells the size of single mitochondria varies considerably (e.g. Bereiter-Hahn, 1990) and intensity modulations are observed along single mitochondria (Bereiter-Hahn and Vöth, 1996); for example, see Fig. 1. The total fluorescence emitted from a single mitochondrion under a given metabolic condition depends on its size.

For the explanation of variations in intramitochondrial fluorescence intensity, two factors must be discussed. First, the intensity variations reveal variations in the amount of mitochondrial inner membranes per unit area (or voxel), and second the intensity variations monitor differences in the electrochemical gradient.

The rough positive correlation between mitochondrial thickness and fluorescence intensity is indicative of variation in inner membrane mass per unit area (Fig. 1). Comparison of the fine structure of single mitochondria as revealed by electron microscopy and fluorescence intensity distribution over the same mitochondrion was used to see whether crista distribution and fluorescence intensity correlate with each other. Examples are given in Figs. 2–5.

Pseudocolor coding shows that the highest relative fluorescence intensities occur in the central parts of the filamentous mitochondria, and, as a first approximation, intensity seems to be related to thickness (compare ends of mitochondria 1, 2, 6, 15 in Fig. 1). Such a relationship also exists in rounded mitochondria; greatest brightness is observed in the large mitochondria numbers

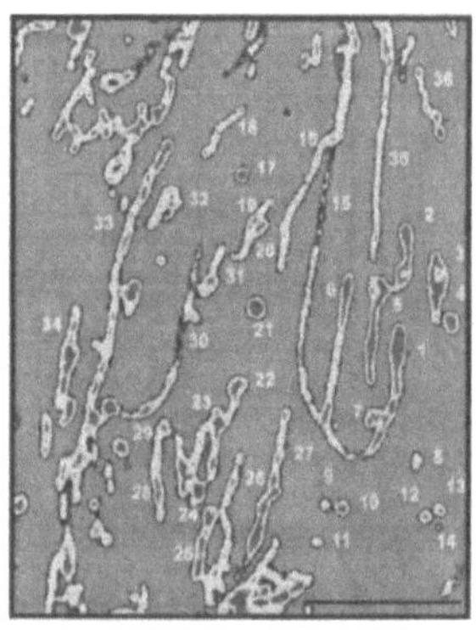

Fig. 1. Mitochondria stained with diamino-styryl-pyridinium methiodide (*DASPMI*) in the periphery of a living XTH-2 cell. Pseudocolouring reveals differences in fluorescence intensity. *Numbers* mark some of the mitochondria that have been revealed by electron microscopy and are shown in Figs. 2–5. *Bar*: 10 µm

Figs. 2–5. Comparison of the vital appearance of single mitochondria after staining with DASPMI (pseudocoloured images) with corresponding electron micrographs. The shapes of the mitochondria in the fluorescence images and in the electron micrographs may differ from each other, because of shape changes taking place in the time interval (up to 30 s) between acquisition of the fluorescence image and stabilization of structure by fixation. Mitochondria are associated with microtubules and endoplasmic reticulum, as can be seen in most electron micrographs

21 and 4, the equally sized mitochondria 12 and 13 exhibit exactly the same fluorescence while the small ones emit less fluorescence (numbers 8, 9, 11, 14). However, the projection area of mitochondrion 8 equals that of 17 but the fluorescence differs significantly. A major problem in relating thickness to fluorescence intensity is that a circular cross-section of the organelles can be assumed only for thin parts of a mitochondrion; in the thicker areas the organelles become flattened because of the very small thickness of the cytoplasm in this region (0.5–1.5 μm). On the other hand, this cell architecture is a prerequisite for correlative microscopy comparing the fluorescence of the living cell with its fine structural details as revealed by electron microscopy.

Crista density does not necessarily correlate with fluorescence intensity

Mitochondrion 10 is almost devoid of cristae (Fig. 3b), the neighbour number 9 contains more and fluoresces less, although both are almost identical in size. Brightly fluorescing mitochondria 4 and 21 (Fig. 1), and the smaller but bright pair 12 and 13 contain cristae at a normal density (electron micrographs not shown). Numerous mitochondria with a speckled fluorescence distribution appear electron translucent in the corresponding electron micrographs (e.g. mitochondria 5 and 6 in Fig. 2; 16, 19 and 20 in Fig. 4; 18 in Fig. 1; 22–28 in Fig. 3; 31 in Fig. 5). Certainly, these crista-free, translucent zones are artefacts caused by photodamage due to the illumination of the mitochondria with the light for fluorescence excitation. However, the fact that photodamage is a localized event points to the limitations of diffusion inside mitochondria. The dye became more concentrated in certain areas and thus more free radicals were released by irradiation. Some of these speckled areas are separated from each other by cristae crossing the whole width of the organelle (arrows in numbers 1 and 5 in Fig. 2; 28 in Fig. 3; 31 in Fig. 5). Mitochondrial areas that are almost devoid of cristae regularly occur along the tapering ends of very long mitochondria; numbers 15 (Fig. 2) and 30 (Fig. 5) are examples of this type of organization. Also in these zones brighter fluorescing spots correspond to electron-translucent areas (i.e. mitochondrion 15 in Fig. 2). In most cases the translucent, damaged areas appear swollen. The local restriction of swelling is a further argument for submitochondrial compartmentation. No correlation of this type of compartmentation with the distribution of mitochondrial nucleoids (see Bereiter-Hahn and Vöth, 1996) was found (results not shown).

JC-1 fluorescence of single mitochondria

Staining of mitochondria in living XTH-2 cells with the dye JC-1 results in a green staining of whole mitochondria and, in addition, some parts of a mitochondrion may appear red, if the fluorescence is excited with blue light (450–490 nm) (Smiley *et al.*, 1991). This red fluorescence was shown to be linearly related to the membrane potential of mitochondria in suspension (Reers

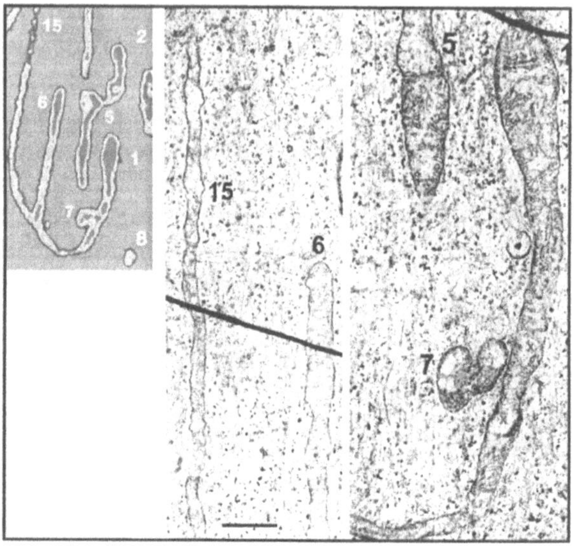

Fig. 2. The fluorescence image did not show the two mitochondria 1 and 7 separately. Two of the limited areas with brighter fluorescence are marked by a *red X* in the electron micrograph of mitochondrion 1. The *arrowheads* next to mitochondria 1 and 5 point to cristae traversing the whole width of the mitochondria. The very thin part of mitochondrion 15 contains very few cristae only

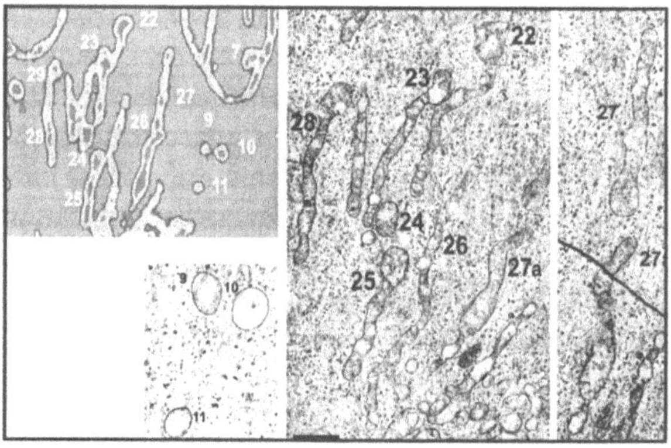

Fig. 3. Group of mitochondria with numerous electron-translucent areas corresponding to zones of bright fluorescence in the pseudocolour images. Two consecutive sections of mitochondrion 27a are shown (for further details see text)

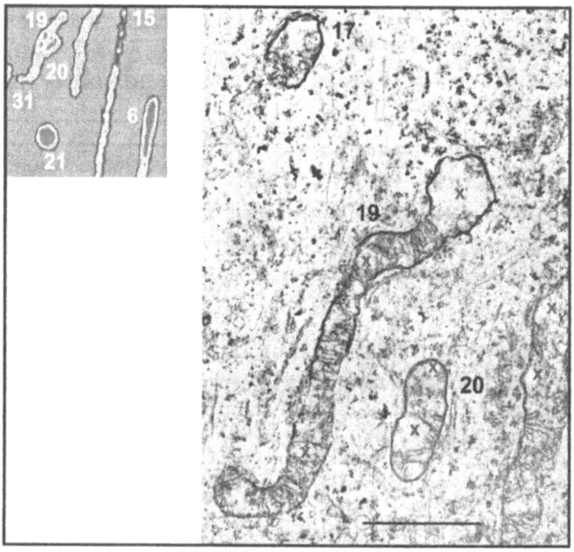

Fig. 4. Electron-translucent areas in these mitochondria corresponding to zones of bright fluorescence in the pseudocolour images are marked by *red X's*. Number 20 has slightly changed its position between fluorescence image acquisition and fixation. Number 16 is shown at two consecutive sections which have been properly aligned

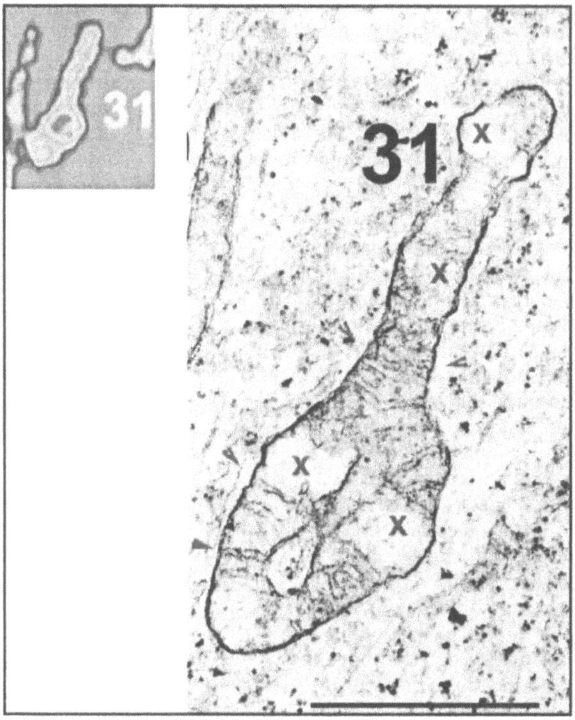

Fig. 5. Further example of colocalization of brighter areas of fluorescence and electron-translucent zones (*red X's*), and some cristae traversing the whole width of a mitochondrion (*arrowheads*)

et al., 1991, 1995). The mosaic pattern of mitochondrial staining may indicate local membrane potential differences along a single organelle. Therefore, we compared the fluorescence pattern of single mitochondria stained while they were alive with their fine structure as revealed from electron micrographs of thin sections after fixation with glutardialdehyde and osmium tetroxide. The observations for six mitochondria are shown in Figs. 6–8. Those areas of mitochondria that contain a large mass of cristae appear red (e.g. head of mitochondrion number 1 in Fig. 6, number 2 and both ends of 3 in Fig. 7, numbers 4 and 6 in Fig. 8). This is most obvious in the dumbbell-shaped mitochondrion 6: the two ends fluoresce red, and in phase-contrast they appear darker than the rest of the mitochondrion. The thick head region of mitochondrion 1 (Fig. 6) fluoresces red strongly, the „nose" does not, although it contains cristae that are indistinguishable in shape, swelling degree and matrix density from the head region. Again, the red fluorescing end of the tail does not differ in structure, but more crista profiles appear in this region than along the rest of the tail. Mitochondria 5 and 6 do not differ in their fine structure (Fig. 8), but only number 6 exhibits red fluorescence.

Figs. 6–8. Comparison of the vital appearance of single mitochondria after staining with JC-1 with corresponding electron micrographs. All mitochondria fluoresce green (not shown), red fluorescence is shown in the *insets* in the phase-contrast micrographs. The shape of mitochondria in the phase-contrast microscopy image may deviate from that in the electron micrographs because of the mitochondria's abilities to change shape very quickly. *Numbers* mark some of the mitochondria which have been retrieved by electron microscopy and are shown in Figs. 6–8. *Bar*: 1 µm

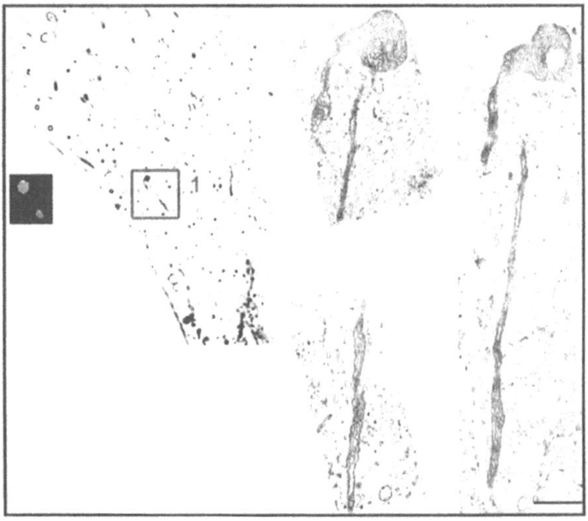

Fig. 6. Mitochondrion number 1 consists of a thick head region and a long and thin tail region. The electron micrographs show two subsequent sections through this mitochondrion. The head region and the distal tail region contain many cristae and a dense matrix whilst the thin region connecting the elongated tail part to the head contains very few inner membranes only. Thus red fluorescence (*inset* in the phase-contrast image) is found in the thick head and the thickened tail end. The middle part of the tail region is devoid of red fluorescence

Bleaching of mitochondrial fluorescence

If mitochondria should manage to organize zones with different membrane potential, local injury caused by a small area of fluorochrome bleaching should not spread along the whole mitochondrion. Such an experiment was designed first by Skulachev and his group (Amchenkova *et al.*, 1988). We repeated this experiment using elongated mitochondria in living endothelial cells (XTH-2 cells). Some mitochondria, fluorochromed with DASPMI, were irradiated with a laser beam (λ=488 nm) crossing the long axis of the mitochondrion at a diffraction-limited line (width approx. 0.3 µm). Immediately after the bleaching pulse (approx. 2–5 s later) a picture of the whole field was taken. The mitochondria crossing the bleaching line (shown in Fig. 9b) became totally decolorized, whilst the fluorescence of other mitochondria in their immediate vicinity did not change (Fig. 9). If the mitochondria was prestained with Mitotracker, a fluorochrome developing its fluorescence on mitochondria-mediated reduction, loss of fluorescence also occurred, but it recovered after a few minutes left in the presence of Mitotracker. These experi-

Fig. 7. The thick mitochondrion number 2 and the thicker ends of mitochondrion number 3 fluoresce red, whilst the thinner centre of the bent mitochondrion, number 3, does not exhibit red fluorescence, although it does not differ in the local arrangement, matrix density and density of cristae

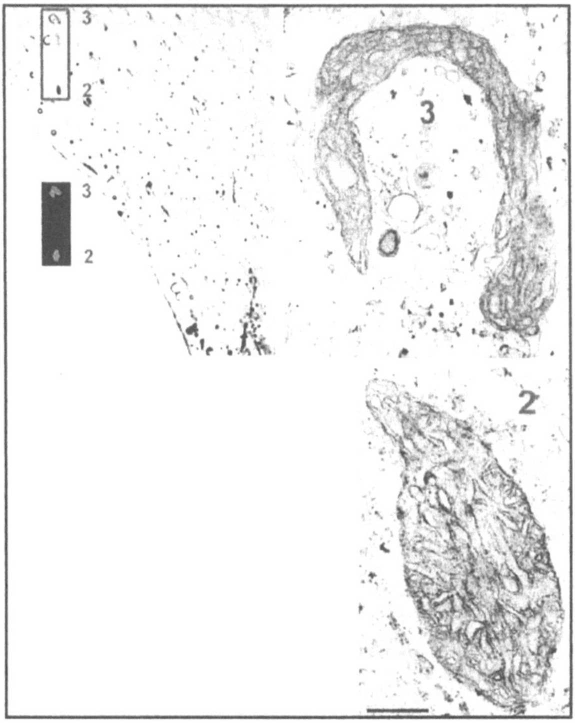

ments argue for a generalized release of dye molecules and of membrane potential as well, through small localized injuries, and do not support the idea of patterned membrane potential. The recovery of the Mitotracker fluorescence shows that the bleached mitochondria retain the ability for electron transport; however, their electrochemical coupling is impaired because DASPMI fluorescence does not recover over the next 15 min.

Discussion

The fluorescence intensity of mitochondria stained with fluorochromes depends on the amount of dye accumulated per unit volume and on the quantum yield of the dye molecules. In two-dimensional images fluorescence does not reveal unequivocally the dye concentration (mass of dye per unit volume), but rather the mass of dye per unit area, if we assume a constant quantum yield inside mitochondria. Therefore, the fluorescence emitted by fluorochromes that become situated in the crista membrane should be related to the density of cristae inthe projection area. Thus, organelle thickness and crista density have to be considered the main factors that determine local fluorescence intensity. A rough correlation between organelle density and fluorescence distribution exists. For the endothelial cells under investigation this cor-

Fig. 8. Three mitochondria (4–6). Number 5 does not fluoresce red although its fine structure equals that of number 6. Number 4 is a bit thicker and has a dense matrix and high crista density. However in both 4 and 6, red fluorescence is found at the opposite ends and not in the connecting middle part. This middle part in 6 does not differ from the ends, as is shown by the series of three subsequent sections through this mitochondrion (pictures at the *bottom* of the *figure*)

relation cannot be established exactly, because the cross-section of a mitochondrion is not known and cannot be determined with sufficient accuracy (the *z*-resolution of confocal laser scanning microscopy does not allow exact measurement of the thickness in the range 0.4–1 μm). Only in the case of very thin mitochondria can a circular cross-section be assumed. However, the electron micrographs show clearly that no alterations in crista density are found in zones exhibiting red J-C1 fluorescence. The very distinct staining pattern with J-C1 along some mitochondria was interpreted as a mosaic pattern of membrane potential distribution (Smiley *et al.*, 1991). This pattern is not paralleled by any localized differences in crista development. The thickness of the mitochondria, i.e. mass of mitochondrial membranes per unit area, may be one of the main factors controlling the transient J-C1 red fluorescence. On the other hand, the fluorescence of the potential-sensitive dye DASPMI also varies along single mitochondria. In isolated mitochondria DASPMI fluorescence follows the Nernst distribution, as revealed by K^+ distribution in nigericin-permeabilized mitochondria. Also this fluorescence, in general, is related to the local thickness of the organelles. However, intensity changes are also found along organelles with almost constant thickness, as was shown for J-C1 by Chen and co-workers (Smiley *et al.*, 1991). These local intensity changes may become displaced along a single mitochondrion, they may appear and disappear (Bereiter-Hahn and Vöth, 1996). These observations again point to the possibility of membrane potential variations inside a single mitochondrion. The potential differences between two adjacent areas in a single mitochondrion have been estimated according to the Nernst distribution at up to 15 mV. Most probably these small variations in potential do not cause visible morphological alterations of the cristae.

Contrary to the rhodamines, DASPMI seems to accumulate in the crista membrane rather than the matrix. It easily partitions from the aqueous medi-

Fig. 9a–c. Confocal laser scanning microscopic images of DASPMI-stained mitochondria in a living XTH2-cell. a Before; b immediately after the laser irradiation along the line indicated; c 100 s after irradiation. Some of the mitochondria have been numbered; they stopped fluorescing along their whole length, whilst some mitochondria in their immediate vicinity remained fluorescing (*arrowheads*). As a consequence of the laser irradiation required for taking the pictures a general decrease of fluorescence and total bleaching of some mitochondria may occur (*arrow*). *Bar*: 10 μm

um into a hydrohobic environment and through this considerably increases its quantum yield. This increase, together with the membrane-potential-driven uptake of the dye, is assumed to explain the selectivity of DASPMI fluorescence for mitochondria in living cells. Accumulation in the cristae rather than the matrix renders the membranes more sensitive to photodamage.

Local variations of membrane potential inside single mitochondria require restricted diffusion of protons along the organelles. The experiments with local photobleaching and the high motility of mitochondria (extensive shape changes) argue against such a restricted intramitochondrial diffusion. However, the disappearance of cristae on illumination of DASPMI-stained mitochondria just in those areas of particularly high fluorescence are a strong argument for high local dye concentrations and, thus, restricted diffusion of the molecules. In unstained mitochondria areas devoid of cristae are never observed (see for example Bereiter-Hahn and Vöth, 1983). The phototoxic effect may also induce some local swelling and thus further indicates the existence of subcompartments along mitochondria. Separation of such subcompartments could be reached by cristae traversing through the whole width of an organelle (e.g. arrows in Figs. 2 and 3, mitochondrion 20 in Fig. 4). These observations can be interpreted as an indication of local variations in membrane potential along a single mitochondrion revealed by staining with DASPMI.

The generalized bleaching along the full length of a mitochondrion by a localized strong bleaching pulse shows that permeabilization at a single site may discharge the membrane potential along the total length of a large mitochondrion. Also, dye molecules accumulated in the matrix that do not depend on the mitochondrial membrane potential (in the case of Mitotracker) can be released within a few seconds. This experiment does not exclude the possibility of the development of local membrane potential variations, because:

1. The bleaching proceeds slowly along some mitochondria (Fig. 9).
2. The net membrane potential results from the steady-state conditions of respiration-driven proton efflux and re-flux through the ATP-synthase.

In case these transmembrane transport mechanisms are faster than lateral diffusion in the membrane or along the mitochondrion in the matrix, local differences in the steady-state conditions can be supposed to occur. Whether these reflect differences in local energy demand in the cytoplasm remains to be investigated.

For the distribution of the JC-1 red fluorescence the situation may be similar in so far as J-aggregate formation starts at sites with the highest local JC-1 concentration. Accumulation of the dye requires actively respiring cristae, and where these have a relatively high density or they develop a higher local membrane potential, the uptake into the matrix may be faster than in other areas. Thus, in thick zones of mitochondria containing many cristae, the critical concentration for J-aggregation may be reached more quickly than in thinner areas with a smaller cristae mass, where the dye diffusion balances that of accumulation. Sites of developing J-aggregates may trap dye molecules from their vicinity and thus a regular pattern of red and green fluorescence may appear (Smiley *et al.*, 1991). This interpretation is in accord with the observation that the red fluorescence pattern in living cells is very unstable with time. Because of its transient character, this type of fluorescence does not seem to be a reliable indicator of mitochondrial membrane potential *in situ*. For isolated mitochondria the situation is different, insofar as these mitochondria no longer maintain their elongated and highly variable shape. Due to the isolation procedure these mitochondria are spherical and of similar size; thus, in isolated mitochondria formation of J-aggregation may be an excellent indicator of the membrane potential.

Acknowledgements. JC-1 was a generous gift of Dr. Lan Bo Chen (Boston) which is gratefully acknowledged. This work has been supported by the Deutsche Forschungsgemeinschaft: SFB 474 (project B5).

References

Allmann, D.W., Munroe, J., Wakabayashi, T., Harris, R.A., Green, D.E. (1970a) Studies on the transition of the cristal membrane from the orthodox to the aggregated configuration. II. Determinants of the orthodox-aggregated transition in adrenal cortex mitochondria. *J. Bioenerg.* **1**:87–107

Allmann, D.W., Wakabayashi, T., Kormann, E.F., Green, D.E. (1970b) Studies on the transition of the cristal membrane from the orthodox to the aggregated configuration. I. Topology of bovine adrenal cortex mitochondria in the orthodox configuration. *J. Bioenerg.* **1**:73–86

Amchenkova, A.A., Bakeeva, L.E., Chentsov, Y.S., Skulachev, V.P., Zorov, D.B. (1988) Coupling membranes as energy-transmitting cables. I. Filamentous mitochondria in fibroblasts and mitochondrial clusters in cardiomyocytes. *J. Cell Biol.* **107**:481–495

Bereiter-Hahn, J. (1976) Dimethylaminostyrylmethylpyridiniumiodine (DASPMI) as a fluorescent probe for mitochondria in situ. *Biochim. Biophys. Acta* **423**:1–14

Bereiter-Hahn, J. (1990) Behavior of mitochondria in the living cell. *Int. Rev. Cytol.* **122**:1–63

Bereiter-Hahn, J., Vöth, M. (1983) Metabolic control of shape and structure of mitochondria in situ. *Biol. Cell* **47**:309–322

Bereiter-Hahn, J., Vöth, M. (1994) Dynamics of mitochondria in living cells: shape changes, dislocations, fusion and fission of mitochondria. *Microsc. Res. Tech.* **27**:198–219

Bereiter-Hahn, J., Vöth, M. (1996) Distribution and dynamics of mitochondrial nucleoids in animal cells in culture. *Exp. Biol. Online* **1**:4

Bereiter-Hahn, J., Seipel, K.-H., Vöth, M., Ploem, J.S. (1983) Fluorimetry of mitochondria in cells vitally stained with DASPMI or rhodamine 6 GO. *Cell Biochem. Funct.* **1**:147–155

Chance, B., Williams, G.R. (1956) The respiratory chain and oxidative phosphorylation. *Adv. Enzymol.* **17**:65–134

Chen, L.B. (1988) Mitochondrial membrane potential in living cells. *Annu. Rev. Biol.* **4**:155–181

Chen, L.B. (1989) Fluorescent labeling of mitochondria. *Methods Cell Biol.* **29**:103–123

Cohen, L.B., Salzberg, B.M. (1978) Optical measurement of membrane potential. *Annu. Rev. Physiol. Biochem. Pharmacol.* **83**:35–88

Eng, J., Lynch, R.M., Balaban, R.S. (1989) Nicotinamide adenine dinucleotide fluorescence spectroscopy and imaging of isolated cardiac myocytes. *Biophys. J.* **55**:621–630

Fuchs, J., Zimmer, G., Bereiter-Hahn, J. (1987) A multiparameter analysis of the perfused rat heart: responses to ischemia, uncouplers and drugs. *Cell Struct. Funct.* **5**:245–253

Hackenbrock, C.R. (1968) Chemical and physical fixation of isolated mitochondria in low energy and high-energy states. *Proc. Natl. Acad. Sci. USA* **61**:598–605

Hackenbrock, C.R. (1981) Lateral diffusion and electron transfer in the mitochondrial inner membrane. *Trends Biochem. Sci.* **6**:151–154

Hackenbrock, C.R., Rehn, T.G., Weinbach, E.C., Lemasters, J.J. (1971) Oxidative phosphorylation and ultrastructural transformation in mitochondria in the intact ascites tumor cell. *J. Cell Biol.* **51**:123–137

Harris, R.A., Williams, C.H., Caldwell, M., Green, D.E. (1965) Energized configuration of heart mitochondria in situ. *Science* **165**:700–703

Harris, R.A., Harris, D.L., Green, D.E. (1968) Effect of *Bordetella* endotoxin upon mitochondrial respiration and energized processes. *Arch. Biochem. Biophys.* **128**:219–230

Harris, R.A., Asbell, M.A., Asai, J., Jolly, W., Green, D.E. (1969) The conformational basis of energy transduction in membrane systems. V. Measurement of the configurational changes by light scattering. *Arch. Biochem. Biophys.* **132**:545–560

Irion, G., Ochsfeld, L., Naujok, A., Zimmermann, H.W. (1993) The concentration jump method. Kinetics of vital staining of mitochondria in HeLa cells. *Histochemistry* **99**:75–83

Johannes, H. (1941) Beiträge zur Vitalfärbung von Pilzmyzelien II. Die Inturbanz der Färbung mit Rhodaminen. *Protoplasma* **36**:181–194

Johnson, L.V., Walsh, M.L., Chen, L.B. (1980) Localization of mitochondria in living cells with rodamine 123. *Proc. Natl. Acad. Sci. USA* **77**:990–994

Koke, J.R., Shug, A.L., Folts, J.D., Bittar, N. (1976) Ultrastructural and physiological changes induced by atractylate in the canine myocardium. *Cytobios* **17**:211–229

Koukl, J.F., Vorbeck, M.L., Martin, A.P. (1997) Mitochondrial three-dimensional form in ascites tumor cells during changes in respiration. *J. Ultrastructr. Res.* **61**:58–165

Leip, O., Breipohl, W., Wegener, A., Augustin, A. (1994) Distribution pattern of viable mitochondria in bovine lens epithelial cells. *Dev. Ophthalmol.* **26**:90–96

McCormick, S. (1990) Fluorescent labelling of Na^+,K^+-ATPase in intact cells by use of a fluorescent derivative of ouabain: salinity and teleost chloride cells. *Cell Tissue Res.* **260**:529–533

Munn, E.A. (1974) The structure of mitochondria. Academic, London, pp. 1–465

Ord, M.J., Smith, R.A. (1982) Correlation of mitochondrial form and function in vivo: microinjection of substrate and nucleotides. *Cell Tissue Res.* **227**:129–137

Penniston, J.T., Harris, R.A., Asai, J., Green, D.E. (1968) The conformational basis of energy transformation in membrane system. I. Conformational changes in mitochondria. *Proc. Natl. Acad. Sci. USA* **59**:624–631

Petit, P.X. (1992) Flow cytometric analysis of Rhodamine-123 fluorescence during modulation of the membrane potential in plant mitochondria. *Plant Physiol.* **98**:279–286

Reers, M., Smith, T.W., Chen, L.B. (1991) J-Aggregate formation of a carbocyanine as a quantitative fluorescent indicator of membrane potential. *Biochemistry* **30**:4480–4486

Reers, M., Smiley, S.T., Mottolahartshorn, C., Chen, A., Lin, M., Chen, L.B. (1995) Mitochondrial membrane potential monitored by JC-1 dye *Mitochondrial Biogen. Genet.* **A260**:406–417

Schlage, W., Kuhn, C., Bereiter-Hahn, J. (1981) Established *Xenopus* tadpole heart endothelium (XTH) cells exhibiting selected properties of primary cells. *Eur. J. Cell Biol.* **24**:342

Smiley, S.T., Reers, M., Mottolahartshorn, C., Lin, M., Chen, A., Smith, T.W., Steele, G.D., Chen, L.B. (1991) Intracellular heterogeneity in mitochondrial membrane potentials revealed by a J-aggregate-forming lipophilic cation JC-1. *Proc. Natl. Acad. Sci. USA* **88**:3671–3675

Sobreira, C., Davidson, M., King, M.P., Miranda, A.F. (1996) Dihydrorhodamine 123 identifies impaired mitochondrial respiratory chain function in cultured cells harboring mitochondrial DNA mutations. *J. Histochem. Cytochem.* **44**:571–579

Thürich, T., Bereiter-Hahn, J., Schneider, M., Zimmer, G. (1998) Cardioprotective effects of dihydrolipoic acid and tocopherol in right heart hypertrophy during oxidative stress. *Arzneimittelforschung* **48**:13–21

Van Itallie, C.M., Kieffer, L.J., Carbone, R., Dannies, P.S. (1990) Estradiol decreases retention of rhodamine 123 fluorescence in GH4C1pituitary tumor cells. *Mol. Endocrinol.* **4**:140–145

Walsh Kinnally, K., Tedeschi, H., Maloff, B.L. (1978) Use of dyes to estimate the electrical potential of the mitochondrial membrane. *Biochemistry* **17**:3419–3428

Williams, C.H., Vail, J., Harris, R.A., Caldwell, M., Green, D.E., Valdivia, E. (1970) Conformational basis of energy transduction in membrane system. VIII. Configurational changes of mitochondria. *Bioenergetics* **1**:147–180

Author index

A

B

C

D

E

F

Keywords

Copyright information

License Agreement and Warranty

MIX
Papier aus verantwortungsvollen Quellen
Paper from responsible sources
FSC® C105338

If you have any concerns about our products,
you can contact us on
ProductSafety@springernature.com

In case Publisher is established outside the EU,
the EU authorized representative is:
Springer Nature Customer Service Center GmbH
Europaplatz 3, 69115 Heidelberg, Germany

Printed by Libri Plureos GmbH
in Hamburg, Germany